Internal Swelling Reactions in Concrete

As critical concrete infrastructure deteriorates, engineers need efficient and reliable techniques to appraise the causes and the extent of deterioration, to evaluate the structural consequences and to select effective management protocols and rehabilitation strategies. This book looks at deterioration caused by internal swelling reaction (ISR) mechanisms in concrete, such as alkali-aggregate reaction, delayed ettringite formation and freeze-thaw cycles.

The book provides accessible and comprehensive coverage of recent work and developments on the most common ISR mechanisms leading to induced expansion and deterioration. It addresses the implications of ISR on different scales (micro, meso and macro), outlines qualitative and quantitative techniques to assess the condition of affected concrete and introduces the multi-level assessment protocol, using advanced microscopic and mechanical techniques, particularly the stiffness damage test and damage rating index, as a reliable approach to appraise ISR-affected infrastructure. Also included is a detailed case study of the Robert-Bourassa Charest Overpass in Quebec.

Internal Swelling Reactions in Concrete: Mechanisms and Condition Assessment is primarily intended for undergraduate and graduate students, as well as academics interested in the field of concrete durability and condition assessment of concrete. It will also be of interest to engineers and infrastructure owners dealing with ISR-related problems.

Internal Swelling Reactions in Concrete

Mechanisms and Condition Assessment

Leandro F. M. Sanchez

CRC Press
Taylor & Francis Group
Boca Raton London New York

CRC Press is an imprint of the
Taylor & Francis Group, an **informa** business

Cover image: Leandro F. M. Sanchez

First edition published 2024
by CRC Press
2385 NW Executive Center Drive, Suite 320, Boca Raton FL 33431

and by CRC Press
4 Park Square, Milton Park, Abingdon, Oxon, OX14 4RN

CRC Press is an imprint of Taylor & Francis Group, LLC

ISBN: 978-1-032-03598-7 (hbk)
ISBN: 978-1-032-03599-4 (pbk)
ISBN: 978-1-003-18815-5 (ebk)

DOI: 10.1201/9781003188155

Typeset in Sabon
by SPi Technologies India Pvt Ltd (Straive)

Life is a challenging and yet different journey for everyone. Nobody knows exactly the main objective(s) of such a journey, and all of us have very distinct interests; however, I believe most of us agree that as long as we do things with passion, we are on the right track.

Born and raised in a country with extreme challenges, I have been taught from my early years that *education* should be amongst my main priorities. My parents have done everything they could so that my sister and I could have the best education possible, both informal (at home) and formal (at school). They made me see how important professors were for society and how these outstanding human beings could positively impact the lives of people. These professionals give much more than they take, teach much more than they learn and share much more than they withhold. Yes, professors change lives! They simply do! From teaching important and specific topics to making your dreams possible and helping create the means, professors make the impossible come true. From my best professors, I learnt that if I kept books as my best friends, I should be ok. And my life went on this way…always with a book as my best friend.

Life was super kind to me on this purpose. I have had the opportunity to have amazing professors in my life, much more than I could ever list herein; however, some of them were not only important to my technical knowledge and background but were crucial to shaping the person I am. My first professors were my parents. My mom taught me *Portuguese*, *English*, *history* and *geography*, while my dad taught me *math* and *physics*. Besides being my superheroes, I had two role models directly at home to follow from the very beginning, and I am so glad for it. Growing up, I met numerous great professors at distinct levels of my formal education, and each of them taught me incredible lessons I still remember nowadays.

When I was 16, I was unsure which path to follow and was losing motivation at school. Yet I met a very important physics professor, Ms. Virginia Simoes Cortez, who was the first person who told me I had talent and should pursue a STEM degree. This dictated the choices I would have made in the years to come. Already at the university, when the future was still unclear to me, I registered myself on a course called Applied Math to Life; in this course, I met the incredible professor Aguinaldo Prandici Ricieri, who taught me, among integrals, derivatives and others math and physics tricks, the love for science and the possibility to achieve my dreams in a research-oriented career, as long as I kept my passion and determination. Upon graduation, uncertain of how the best manner would be to formally initiate a graduate programme, I had the opportunity to meet Professor Cardoso, dean of the Polytechnic School of Sao Paulo at the time. Professor Cardoso was one of the best mentors I have ever had in my life. He explained to me, little by little, the career of a researcher and pointed me to the graduate programme at the Polytechnic School of Sao Paulo, where everything started.

I then initiated a *master of applied sciences (MASc)* programme under the supervision of Professors Paulo Helene and Selmo Kuperman; after a preliminary proposal to conduct a project on steel corrosion, for some reason beyond my control, they changed completely my project to work on an international interlab led by a Canadian Institution (CANMET) to study internal swelling reactions (ISRs) in concrete, particularly alkali-aggregate reaction (AAR); I did not know at the time, but I was about to fall in love with ISR and this new project would change my entire career, for the better. Moreover, Professors Paulo Helene and Selmo Kuperman became much more than merely supervisors to me but my career inspirations, mentors and very close friends. Yet during my MASc, I had the opportunity to meet and discuss with so many great professors, and amongst them professor Maria Alba Cincotto, who not only taught me *Cement Chemistry* but also unforgettable lessons on determination, science and ethics, lessons that I carry with me until today. From time to time, I remember her great lessons (inside and outside the classroom) on the Polytechnic School campus in Sao Paulo, and I still miss them.

When my MASc degree was successfully completed, I knew I wanted more, but I could not imagine that the best was yet to come. It was then when life played a very important role, and I received a formal contact from Dr. Benoit Fournier, who had recently quit CANMET and became a professor at Laval University, to pursue a PhD under him on AAR. I promptly accepted the invitation and took it not only as a likely great experience but as a dream coming true and the opportunity of my life. Over the years as a PhD student, I realized very quickly that research, especially on ISR, was exactly what I wanted for my life. Dr. Benoit Fournier, definitely amongst the best researchers in the field, was not only a great supervisor for me but also a wonderful mentor, and why not a *father*? I am so glad and thankful for all the lessons I learnt from him and will never forget all the help and support I received. I remember thinking when I started my position at the University of Ottawa that if I could be half of what Benoit was for me to my students, I would be extremely satisfied!

Towards the end of my PhD, I was introduced to professors Denis Mitchell and Jose Bastien, who finally supervised and guided me in a postdoctoral programme. They also were so important to me, not only on the research programme itself but on everything else that an early career researcher needs, such as support, guidance and motivation. I will never forget some of the discussions we had at McGill. I also miss this time.

Finally, I would like to dedicate this book to my mentor Ramon, along with my beloved Diana (wife), Ian and Noah (kids), who are my daily professors in the art of love...before knowing you, I could never imagine love could be so huge.

This book is dedicated to the professionals who change lives, the ones who take care more of others than themselves, to all professors, particularly to the professors of my life.

Leandro F. M. Sanchez
July 23, 2023
Ottawa, ON, Canada

"In the end, I don't really know whether this work belongs to me, although I'm pretty sure that I completely belong to it".

Leandro F. M. Sanchez

Contents

About the Author

Dr. Leandro F. M. Sanchez obtained a bachelor's and MASc in civil engineering from Maua School of Engineering (Sao Caetano, Brazil, 2004) and Polytechnic of Sao Paulo (Sao Paulo, Brazil, 2008), respectively; a PhD in earth sciences from Université Laval (Quebec City, Canada, 2014); and a postdoctoral fellowship in civil engineering from McGill University (Montreal, Canada, 2015). He joined the Department of Civil Engineering at the University of Ottawa in 2015, where he currently is an associate professor. His expertise is related to the development of sustainable concrete materials and concrete durability, especially the diagnosis and prognosis of concrete affected by internal swelling reactions (ISRs). He has co-authored over 150 refereed publications on concrete sustainability and durability and contributed to the development of standard test protocols, along with descriptive, empirical and numerical models for the design and assessment of concrete. Dr. Sanchez is an active participant on several national and international committees and is currently the deputy chair of the RILEM TC 300 -ARM: Alkali-reaction mitigation in concrete and the chair of ACI Committee 221 – Aggregates. He is also the current chair of the 17th International Conference on Alkali-Aggregate Reaction in Concrete (ICAAR), to be held in Ottawa in 2024. Dr. Sanchez is a reviewer for a number of important journals in the field and is a recipient of highly prestigious awards such as the VANIER PhD scholarship (2010), NSERC Early Career Researcher Award-Discovery Launch Supplement (2016), NFRF-Exploration grant (2019), the distinguished Ontario Ministry of Colleges and Universities Early Career Researcher Award (ERA, 2021) and the Faculty of Engineering Early Career Researcher Award (ECRA, 2023).

Foreword

Professor Leandro F. M. Sanchez presents to the scientific and academic community an extremely important publication on internal swelling reactions (ISR), their mechanisms and how to assess the condition of the ISR-affected concrete structures.

There are currently, in the world, thousands of concrete structures that display deterioration signs at different levels caused by ISR, such as alkali-aggregate reaction (AAR) and internal sulphate attack (ISA); the latter may be caused by either delayed ettringite formation (DEF) or by use of sulphide-bearing aggregates in concrete.

This book covers in detail the most common ISR mechanisms affecting concrete infrastructure worldwide, such as AAR, DEF and freeze-thaw (FT) cycles, which can be particularly a problem in cold climate countries, such as Canada (Chapters 1 and 2). With a very didactic scheme, the author then approaches conventional and advanced techniques used to assess the condition of ISR-affected structures, helping the reader to better diagnose and make predictions (i.e., prognose) of affected structures (Chapter 3). Techniques such as visual inspection (VI) and non-destructive testing (NDT), which are very useful for anyone performing infrastructure evaluations, are comprehensively covered in Chapter 4. Furthermore, two chapters of interest for students and engineers – namely, microscopic analyses and mechanical tests – are thoroughly described in Chapters 5 and 6, emphasizing the various microscopic techniques currently available for diagnosing ISR in concrete along with assisting the physical integrity appraisal of affected concrete, respectively. The multi-level assessment protocol is then presented in Chapter 7, encompassing selected microscopic and mechanical tools earlier discussed and displaying a novel and promising approach to detect the cause and extent of ISR-induced deterioration in concrete.

The sequence of the book brings a very important chapter (Chapter 8 – forecasting future behaviour and managing critical infrastructure affected by ISR), which by far is the least understood topic covered in the book by the ISR community and the one requiring further research and developments; this chapter will greatly help engineers and infrastructure owners. Yet, perhaps the most important chapter in this book is Chapter 9, where a study case is presented by Professor Sanchez, in which a clear understanding of the impact of ISR, particularly AAR and FT, on the durability and long-term performance of a deteriorated overpass in Quebec City, Canada, is achieved via the use of the multi-level assessment. This chapter clearly emphasizes the suitability of the proposed approach to recognize the cause(s) and extent of deterioration of affected structures and support infrastructure owners to make informed decisions on their assets.

Finally, Professor Sanchez, with many years of experience in dealing with cases and studies on internal swelling reactions, points to the future, indicating recommendations and further studies yet to be performed for a better understanding and management of ISR in concrete.

Selmo Chapira Kuperman
Sao Paulo, SP, Brazil

Preface

Critical infrastructure is crucial to society, driving the economy, connecting businesses, communities and people and improving the quality of life of human beings. Nowadays, it is recognized that ISRs are amongst the most harmful deterioration mechanisms affecting the durability, serviceability and long-term performance of critical concrete infrastructure around the globe.

There is currently a large amount of research conducted over the last decades on the most common ISR mechanisms, such as AAR and DEF; however, the available documents bear two important drawbacks: (a) they normally focus on *a single ISR mechanism*, and thus the information on the impact of various ISR mechanisms on concrete is quite spread out, difficult to find and compare to one another; furthermore, the assessment of <u>combined mechanisms</u>, which is very likely to occur in practice, is generally not addressed. And (b) most of these works were developed through a *single "scale" approach*; this means that either microscale (e.g., microscopy), mesoscale (e.g., mechanical tests) or even macroscale (i.e., structural behaviour) is addressed in most of the available documentation. This makes the knowledge developments in this field super deep and specific, which is scientifically outstanding on the one hand but makes the overall understanding of the state of the art in the area quite complicated. The idea of this book was then to develop a single, accessible and comprehensive document in which the recent works and developments on the most common ISR mechanisms leading to induced expansion and deterioration could be discussed; moreover, the implications of ISRs on different scales (i.e., micro, meso and macro) are also addressed, along with discussions on the various stages of the condition assessment of ISR-affected structures, from preliminary (i.e., diagnosis), to middle (i.e., prognosis) or even final stages (i.e., rehabilitation strategies), which can help infrastructure owners and engineers to better cope with ISR-induced deterioration in practice. This book is primarily intended for undergraduate and graduate students and academics interested in the field of concrete durability and condition assessment of concrete as well as engineers and infrastructure owners dealing with problems related to ISRs. Hopefully, this book will be valuable and will support engineers and owners to make better and more informed decisions on the management of ISR-deteriorated infrastructure assets.

Leandro F. M. Sanchez
Ottawa, ON, Canada

Acknowledgments

I was hired in July 2015 to work in the Department of Civil Engineering at the University of Ottawa. Over the last eight years, my life has never been so busy and hectic, and yet I have never had so much fun! We created a new research group (i.e., μStructure) that works on innovative and challenging theoretical and applied research related to concrete technology, sustainability and durability. This group currently comprises over 25 students among volunteers, undergrads, MAScs, PhDs and postdoctoral fellows. It is composed of people with distinct backgrounds and interests, bearing high technical skills, ethics and values.

As time went by, our relationship became above and beyond research and academia, and we actually developed a "concrete" family to which we all feel we belong. As I always tell them, it is definitely not about "me" but rather about "us"; it is always because of the group and for the group. This book would not be possible without the extremely important help and support I have had from several of my super bright students. From drafting and reading texts, finding issues and correcting typos, to working on figures, tables and references, these people were crucial for this project to take place. My special thanks and gratitude to my very special, passionate and talented students: Cassandra Trottier (PhD student), Ana Bergmann (PhD student), Rennan Medeiros (PhD student), Diego Jesus de Souza (former PhD student), Andisheh Zahedi (former PhD student) and Thuc Nguyen (former PhD student). You all rocked it and were side by side with me from the beginning to the end of this project. I am so glad and lucky to have all of you in our family and will never forget what you have done!

Leandro F. M. Sanchez
Ottawa, ON, Canada

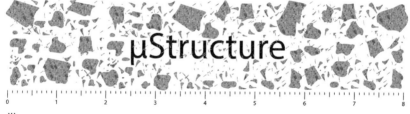

Chapter 1

Introduction

1.1 INTRODUCTION

Our lives critically depend on concrete infrastructure on an everyday basis. Such infrastructure is designed with a lifespan of 50–75 years according to the code used and the structure type (i.e., dams, bridges, tunnels, buildings, etc.). A large number of critical structures built in the 1960s to 1980s worldwide are now reaching the end of their service lives, besides presenting major distress signs caused by numerous damage mechanisms. Action is then required to ensure adequate performance over their last few years of service or even to extend their lifespan.

Amongst the main processes affecting critical concrete infrastructure around the globe, *internal swelling reactions* (ISRs) are likely the most harmful mechanisms, leading to induced expansion, mechanical properties and physical integrity reductions of affected structures and structural members, and decreasing their durability and serviceability performance. ISR is conventionally associated with *alkali-aggregate reaction* (AAR) and *delayed ettringite formation* (DEF), although other mechanisms, such as freeze and thaw and internal sulphate attack derived from sulphide-bearing aggregates, may also induce internal swelling processes leading to induced expansion and deterioration (Sanchez et al., 2018; Noël et al., 2018).

AAR is normally divided into alkali-silica reaction (ASR) and alkali-carbonate reaction (ACR); ASR is by far the most common process affecting concrete structures around the world (Fournier & Bérubé, 2000). ASR is a chemical reaction between the alkali hydroxides (i.e., Na^+, K^+, OH^-) from the concrete pore solution and certain unstable mineral phases found in the aggregates used to make concrete; ASR produces a secondary reaction product (i.e., ASR gel) that swells upon moisture uptake, leading to induced expansion and cracking of the affected material (Fournier & Bérubé, 2000). Otherwise, ACR is a much less common reaction whose distress mechanism is still mostly unknown, being considered as a form of ASR by some authors (Katayama, 2010; Katayama & Grattan-Bellew, 2012) while other researchers believe that ACR follows a "different" and unique mechanism (Canada Standards Association [CSA Group], 2019 – Appendix B). Nevertheless,

ACR is assumed to take place with the use of carbonate rocks in concrete through a process called "dedolomitization" with the formation of brucite and calcite (Fournier & Bérubé, 2000).

DEF is defined as the formation of ettringite in concrete after setting (or at least a substantial portion of the whole setting process) without the penetration or diffusion of any supplementary external sulphate source from the surroundings (Taylor et al., 2001; Collepardi, 2003; Martin, 2011). Usually, DEF takes place whether the concrete undergoes temperatures above 65°C in the very early hours after pouring, which is very common in either steam-cured members at high temperatures or in mass concrete due to the temperature rise during cement hydration (Taylor et al., 2001; Collepardi, 2003; Martin, 2011). Upon high relative humidity over continuous time periods, DEF leads to induced swelling and deterioration of affected concrete (Taylor et al., 2001; Collepardi, 2003; Martin, 2011).

Over the last decades, several approaches and recommendations, including a variety of laboratory test procedures, have been developed worldwide to assess the "potential" of ISR-induced expansion and deterioration in concrete, along with the efficiency of preventive measures before concrete pouring. Despite some issues with some of these test procedures, the majority of experts agree that, in general, it is now possible to build new concrete infrastructure with minimum or calculated risk of ISR. However, there is currently no consensus about the most efficient method(s) that should be implemented and when for the rehabilitation of existing concrete structures and structural members affected by ISR. In this context, numerous engineers and researchers have been proposing appraisal techniques that are able to determine both the damage's cause and extent (i.e., *diagnosis*) and the potential of further distress (i.e., *prognosis*) of ISR-affected concrete, which are essential steps in selecting efficient rehabilitation methods and optimum application periods for affected concrete infrastructure.

The Institut Français des Sciences et Technologies des Transports, de l'Aménagement et des Réseaux (Ifsttar 2003; now Gustave Eiffel University) Bérubé et al. (2005), Fournier et al. (2010) and Godart et al., 2013 have developed comprehensive management protocols for the diagnosis and prognosis of ISR-affected concrete structures. The proposed protocols are based upon a series of comparative field and laboratory investigations to confirm that ISR is the main cause (or a significant contributor to the overall deterioration observed), thus aiming to select appropriate rehabilitation strategies. Such investigations include one or several of the following steps illustrated in Figure 1.1 and described next (Fasseu & Mahut, 2003; Bérubé et al., 2005; Fournier et al., 2010; Godart et al., 2013):

- Routine field inspection of the structure under study to identify the presence, distribution and severity of the defects affecting the various structural members (especially the ISR-related features), as well as the exposure conditions to which the structure is subjected.

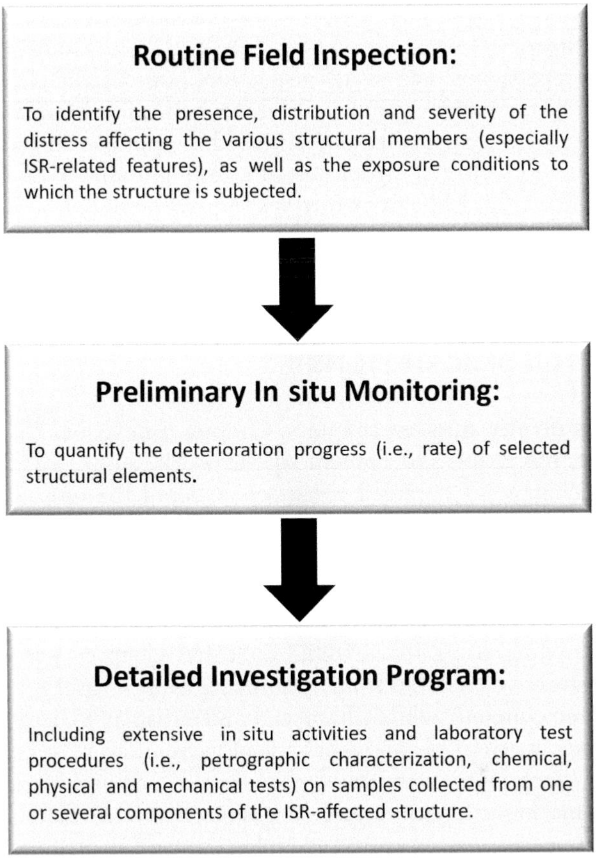

Routine Field Inspection:

To identify the presence, distribution and severity of the distress affecting the various structural members (especially ISR-related features), as well as the exposure conditions to which the structure is subjected.

Preliminary In situ Monitoring:

To quantify the deterioration progress (i.e., rate) of selected structural elements.

Detailed Investigation Program:

Including extensive in situ activities and laboratory test procedures (i.e., petrographic characterization, chemical, physical and mechanical tests) on samples collected from one or several components of the ISR-affected structure.

Figure 1.1 General flowchart for management protocols of ISR-affected structures.

- Preliminary in situ monitoring programme of deterioration (especially signs of induced expansion and deformation) to quantify the progress (i.e., rate) of deterioration on selected structural elements.
- Whenever appropriate (depending on the nature of deterioration and importance of the structure), the implementation of a detailed investigation programme including extensive in situ activities and laboratory test procedures (i.e., petrographic characterization, chemical, physical and mechanical tests) on samples collected from one or several components of the ISR-affected concrete structure.

Although the above protocols and respective steps are very detailed and have been published for quite some time (i.e., 8–16 years), there is still a lack of general understanding of the use of qualitative and quantitative in situ and/or laboratory procedures to appraise the current and future condition of ISR-affected concrete. The goal of the present book is thus to thoroughly describe the most common ISR mechanisms leading to induced expansion and deterioration along with presenting the state of the art of the distinct approaches (i.e., visual inspection, non-destructive tests, chemical procedures, microscopic analyses and mechanical techniques) commonly used or with the potential to be implemented to appraise current and future conditions of ISR-affected structures; the ultimate aim of this book is to provide engineers and infrastructure owners with tools for making better decisions and selecting more appropriate strategies to cope with ISR in critical infrastructure/assets.

1.2 CHAPTER DESCRIPTIONS

This book is divided into ten chapters. Chapter 1 describes the importance of the book, the scope and content of the work, along with the description of the chapters. Chapter 2 discusses in detail the most common ISR mechanisms inducing expansion and deterioration in concrete. Chapter 3 presents the common practices to assess the condition in ageing concrete infrastructure, introduces ISR as an ongoing damage mechanism in concrete, establishes the definition of damage within the context of this work and discusses diagnosing ISR-affected concrete. Chapter 4 presents visual and non-destructive tests (NDT) as promising tools to assess the condition of ISR-affected concrete, while Chapters 5 and 6 display in detail established and promising microscopic and mechanical techniques, respectively.

Chapter 7 introduces the multi-level assessment, a combination of microscopic and mechanical techniques, to quantitatively assess the cause and extent of damage in concrete affected by ISR, as well as discusses the impact of different ISR mechanisms on the engineering properties and structural responses of affected concrete members.

Chapter 8 presents the most common prognosis techniques and management protocols used worldwide and discusses their positive aspects and limitations. Chapter 9 displays a detailed case study from a Canadian overpass affected by ISR after nearly 50 years of service where the proposed multi-level assessment has been successfully implemented. Finally, Chapter 10 presents the conclusions of the book and discusses the need for future work in the field.

REFERENCES

Bérubé, M.-A., Smaoui, N., Bissonnette, B., & Fournier, B. (2005). Outil d'évaluation et de Gestion Des Ouvrages d'art Affectés de Réactions Alcalis-Silice (RAS). http://www.mtq.gouv.qc.ca/portal/page/portal/Librairie/Publications/fr/ministere/recherche/etudes/rtq0608.pdf
Canada Standards Association (CSA Group). (2019). CSA A23.1:19/CSA A23.2:19.

Collepardi, M. (2003). A state-of-the-art review on delayed ettringite attack on concrete. *Cement and Concrete Composites*, *25*(4-5 SPEC), 401–407. https://doi.org/10.1016/S0958-9465(02)00080-X

Fasseu, P., & Mahut, B. (2003). Aide à La Gestion Des Ouvrages Atteints de Réactions de Gonflement Interne. *Guide Technique Des LPC (LCPC, 2003)*.

Fournier, B., & Bérubé, M. A. (2000). Alkali-aggregate reaction in concrete: A review of basic concepts and engineering implications. *Canadian Journal of Civil Engineering*, *27*(2), 167–191. https://doi.org/10.1139/l99-072

Fournier, B., Bérubé, M.-A., Folliard, K. J., & Thomas, M. (2010). Report on Diagnosis, Prognosis and Mitigation of ASR in Transportation Structures. *Federal Highway Administration Publications FHWA-HRT-04-113 (2004) and Techbrief FHWA-HRT-06-071*, issued 2010.

Godart, B., De Rooij, M., & Wood, J. G. M. (2013). *Guide to Diagnosis and Appraisal of AAR Damage to Concrete in Structures – Part 1- Diagnosis*. RILEM State-of the-Art Report (STAR), Springer, 2013.

Katayama, T., & Grattan-Bellew, P. E. (2012). Petrography of Kingston Experimental Sidewalk at Age 22 Years. ASR as the Cause of Deleteriously Expansive, so-Called Alkali-Carbonate Reaction. In *14th ICAAR - International Conference on Alkali-Aggregate Reaction in Concrete*.

Katayama, T. (2010). The So-Called Alkali-Carbonate Reaction (ACR) - Its mineralogical and geochemical details, with special reference to ASR. *Cement and Concrete Research*, *40*(4), 643–675. https://doi.org/10.1016/j.cemconres.2009.09.020

Martin, R.-P. (2011). Analyse Sur Structures Modèles Des Effets Mécaniques de La Réaction Sulfatique Interne Du Béton.

Noël, M., Sanchez, L., & Tawil, D. (2018). Structural implications of internal swelling reactions in concrete: Review and research needs. *Magazine of Concrete Research*, *70*(20), 1052–1063. https://doi.org/10.1680/jmacr.17.00383

Sanchez, L.F.M., Drimalas, T., Fournier, B., Mitchell, D., & Bastien, J. (2018). Comprehensive damage assessment in concrete affected by different Internal Swelling Reaction (ISR) mechanisms. *Cement and Concrete Research*, *107*. https://doi.org/10.1016/j.cemconres.2018.02.017

Taylor, H. F.W., Famy, C., & Scrivener, K. L. (2001). Delayed Ettringite Formation. *Cement and Concrete Research 31*(5), 683–693. https://doi.org/10.1016/S0008-8846(01)00466-5

Chapter 2

Internal swelling reactions (ISRs) mechanisms

2.1 INTRODUCTION: INTERNAL SWELLING REACTIONS (ISRS)

ISRs are deterioration processes leading to induced expansion and damage of affected concrete, usually in the presence of moisture. ISRs are generally associated with a reduction in the mechanical properties, durability and long-term performance of affected concrete (Fournier & Bérubé, 2000; Martin, 2010; Sanchez, 2014). Among existing ISRs, alkali-silica reaction (ASR) and delayed ettringite formation (DEF) are likely the most observed worldwide (Sanchez et al., 2018). However, other mechanisms inducing expansion and damage, such as freeze and thaw cycles (FT), internal sulphate attack (ISA), thaumasite formation (TF), physical sulphate attack (PSA) and hydration of crystalline MgO and CaO, might also be classified as ISR.

2.2 ALKALI-AGGREGATE REACTION (AAR)

AAR is a term used to describe chemical reactions between certain mineral phases from the aggregates used in concrete (fine and coarse) and the alkali hydroxides (i.e., Na^+, K^+ and OH^-) from the concrete pore solution. The first recorded cases of AAR date from the early 1940s in California (United States) by Thomas Stanton (1941); afterwards, numerous cases have been identified worldwide. Nowadays, AAR is recognized as one of the most harmful processes affecting the durability and long-term performance of concrete infrastructure in the field.

AAR-induced expansion and damage in concrete are quite heterogeneous. The reaction kinetics (i.e., induction period, expansion rate and ultimate expansion) depends upon several parameters, such as temperature, alkali-loading of the concrete, type and nature (i.e., particle size and mineralogy) of aggregates, presence of moisture and stress-state. Nevertheless, three conditions are simultaneously necessary to trigger AAR in concrete (Figure 2.1): (a) the presence of reactive aggregates in the mixture, (b) high concentration

DOI: 10.1201/9781003188155-2

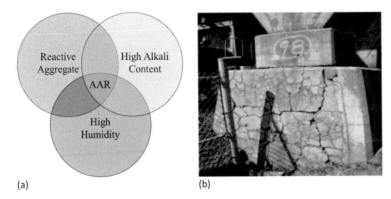

(a)   (b)

Figure 2.1   (a) Three conditions simultaneously required to trigger AAR. (b) Concrete member from Robert-Bourassa/Charest overpass built in 1966 in Quebec City, Canada, using an alkali-silica reactive coarse limestone aggregate.

of alkali hydroxides (i.e., NaOH, KOH) in the concrete pore solution and (c) high humidity conditions. AAR often leads to extensive cracking, loss of material's physical integrity and, in some cases, the functionality of affected structures or structural members.

2.2.1 Induced expansion mechanism

AAR can be divided into two main reaction types: ASR and alkali-carbonate reaction (ACR). ASR is by far the most common reaction type found around the world, and its distress mechanism is already well understood, at least in its major steps. It involves a chemical reaction between the alkali hydroxides from the concrete pore solution and poorly crystalline or metastable silica mineral forms in natural or synthetic aggregates. The reaction generates a secondary product (the so-called *ASR-gel*) that induces expansive pressures within the reacting aggregate material(s) and the adjacent cement paste upon moisture uptake (Fournier & Bérubé, 2000). ASR-induced development can be divided into three major steps (Rajabipour et al., 2015): (a) interaction of the alkaline environment and dissolution of the metastable siliceous material, (b) formation and gelation of colloidal silica and (c) osmotic moisture absorption and secondary products (i.e., ASR-gel) expansion, as presented in Equation 2.1.

$$\left(SiO_2\right)_{solid} \xrightarrow{a} \left(SiO_2\right)_{aqueous} \xrightarrow{b} \left(SiO_2\right)_{gel} \xrightarrow{c} \text{swelling of gel} \tag{2.1}$$

When metastable silica encounters the alkaline solution from the concrete pore, a layer of cations (i.e., calcium, potassium and/or sodium) develops on its surface; such a layer tends to have a higher concentration of calcium, which contributes to delaying the release of amorphous silica, while the diffusion of alkali ions is not affected in the same magnitude. Nevertheless, the metastable silica gradually breaks down and starts bonding with the soluble alkali ions. The resulting product within the aggregate particles bears a somewhat uniform composition, often characterized by a high Si/Ca ratio and high concentrations of Na^+ and K^+ (Rodrigue et al., 2020). However, once cracks extend from the reactive particles towards the cement paste, the reaction products uptake calcium ions encountered at locations close to the interface between the aggregates and the cement paste, which makes the composition of reaction products change with an increase in calcium content, stiffness, viscosity, and swelling potential (Leemann & Lura, 2013).

On the other hand, ACR is a much less common deterioration process whose mechanism is still mostly unknown. ACR is viewed by many researchers as a reaction that occurs between the alkali hydroxides and certain types of dolomitic limestones. However, the threshold between ACR and ASR is unclear in most of the studies. It is generally agreed that ACR is accompanied by the process of dedolomitization and formation of calcite and brucite, displayed in Equation 2.2. This reaction reduces the volume of solids by approximately 13% (Thomas & Folliard, 2007); therefore, induced expansion is attributed to a parallel mechanism. Theories have been proposed to explain the process of ACR-induced expansion, such as (a) hydraulic pressure caused by the migration of water molecules and alkaline ions, (b) absorption of alkaline ions and water molecules on the surfaces of active clay materials scattered by dolomite grains and (c) developments and rearrangement of dedolomitization products and formation and growth of crystalline products in confined spaces. However, it has been claimed that ASR might also play a role in ACR-induced expansion and deterioration in concrete; petrographic examination made on ACR-affected concrete specimens demonstrated the presence of deleterious ASR-gel from cryptocrystalline quartz along with harmless dedolomitization products not associated with induced expansion and crack formation (Katayama & Grattan-Bellew, 2012; Katayama, 2010). Although the aforementioned studies suggest a critical role of ASR in the so-called ACR, the induced expansion and deterioration caused by ACR are significantly different from ASR in affected concrete (i.e., higher expansion rate, mechanical property losses and different crack pattern) as verified by Sanchez et al. (2015).

$$\underbrace{CaMg(CO_3)_2}_{\text{Dolomite}} + 2[\text{Na or K}]OH \rightarrow \underbrace{CaCO_3}_{\text{Calcite}}$$
$$+ \underbrace{Mg(OH)_2}_{\text{Brucite}} + [\text{Na or K}]_2 CO_3$$

(2.2)

2.2.2 AAR microscopic damage features in concrete

AAR damage development is often directly correlated to the level of induced expansion triggered by the physicochemical mechanism. Yet, important progress has been made over the last decades to better understand and explain the impact of AAR-induced microscopic damage features on the mechanical properties of affected concrete.

ASR-secondary products' location and morphology depend on the mineralogical nature of the aggregates (Dunant & Scrivener, 2010). According to the authors, two large classes of aggregate types can be distinguished: (a) slowly reactive and (b) rapidly reactive aggregates. The distress caused by slowly reactive aggregates, which are often used in conventional concrete, is characterized by the formation of secondary products within the aggregates. This phenomenon induces cracks inside the aggregate particles, which reach the bulk cement paste with the increase in expansion. On the other hand, rapidly reactive aggregates are more homogeneous in composition than slowly reactive aggregates where ASR is mainly produced on the surface of their particles; this results in cracks forming in the outer part of the aggregate particles, which leads to cracking (and thus further damage) in the bulk cement paste at early reaction levels (Giaccio et al., 2008). Bérard and Roux (1986) suggested the existence of three types of ASR-induced damage mechanisms as per distinct reactive rock types encountered in Quebec, Canada. These types are illustrated in Figure 2.2:

- Peripheral reactions of non-porous aggregates are shown in Figure 2.2a.
- Diffuse reactions cause the swelling of the bulk reactive aggregate particles, as shown in Figure 2.2b.
- Internal reactions lead to the formation of veins of ASR-secondary products, as displayed in Figure 2.2c.

According to Reinhardt and Mielich (2011), two different distress mechanisms are proposed for ASR-induced development. The first mechanism suggests that the dissolution process takes place at the aggregate particles'

(a)   (b)   (c)

Figure 2.2 AAR-induced damage mechanisms as described by (Bérard & Roux, 1986): (a) peripheral reactions, (b) diffuse reactions and (c) internal reactions.

surfaces; thus, ASR-secondary products and associated cracks are formed at the interfacial transition zone (ITZ) and easily reach the bulk cement paste upon swelling. The second approach states that cracks are generated within the aggregate particles, reaching the cement paste with the increase in induced expansion. This approach assumes that the critical aggregate expansion must be achieved before new cracks are generated in the system; moreover, a critical "crack length" should also be reached before cracks are propagated within the aggregate particles, reaching the cement paste at later stages. Furthermore, the ITZ is observed to remain intact in most cases during ASR development, except in locations close to radial cracks formed within the particles per Golterman (1995). Figure 2.3 illustrates common ASR-induced damage features.

Further investigations made by Sanchez et al. (2015) with a wide range of reactive aggregate types (i.e., fine and coarse aggregates) and natures (i.e., lithotypes), as well as concrete mixtures (i.e., 25, 35 and 45 MPa), demonstrated that at the beginning of the physicochemical reaction (e.g., low expansion levels up to 0.05%), ASR-induced cracks are mainly generated within the aggregate particles. As the expansion progresses and for moderate expansion levels (e.g., ± 0.12%), new cracks are still developed within the aggregate particles; yet, the pre-existing cracks keep growing in length and width, with some cracks reaching the cement paste. At high expansion levels (e.g., ± 0.20%), most of the previously generated cracks may already be found in the cement paste. Furthermore, from this stage and onwards, due to the "minimum energy law", it is easier for ASR-induced expansion to keep increasing pre-existing cracks than generating new cracks in the system. Finally, at very high expansion levels (e.g., ≥0.30%), the cracks previously formed throughout the whole physicochemical process connect to one another in the cement paste, forming an important cracking network that directly impacts the mechanical properties of the affected concrete, especially its compressive strength. Figure 2.4 illustrates the qualitative and

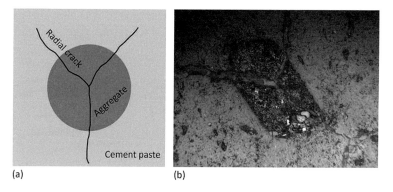

(a)   (b)

Figure 2.3 ASR-induced damage features: (a) scheme with a single aggregate particle and (b) affected concrete specimen (Golterman, 1995).

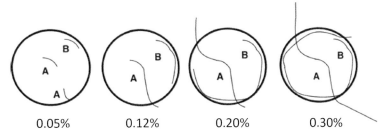

Figure 2.4 Qualitative damage model based on distinct levels of expansion (Sanchez et al., 2015).

descriptive damage model proposed by Sanchez et al. (2015) and described earlier.

Conversely, ACR-induced expansion and damage development were found to be quite different from ASR, at least in its microscopic damage features (Sanchez et al., 2017). Furthermore, although there is no model proposed to explain ACR-induced crack development (i.e., cracks generation and propagation), it has been verified that very few cracks are observed within the aggregate particles at low and moderate expansion levels. In contrast, important cracks are already verified in the cement paste from the beginning of the physicochemical process. Moreover, minimal amounts of reaction products are observed, even at high expansion levels. Therefore, from an engineering and performance point of view, ACR should be considered as a different mechanism when compared to ASR, causing a distinct impact on the mechanical response of affected concrete.

2.2.3 AAR influence on mechanical properties of affected concrete

Evaluating AAR-induced expansion and microscopic damage features discussed in Section 2.2.2 helps us to understand their associated impact on the mechanical properties of the affected concrete. Although conventional mechanical test procedures (i.e., compressive and tensile strengths, modulus of elasticity (ME), etc.) are often used to assess the condition of concrete structures affected by AAR, it is important to discuss the diagnostic character of each of them prior to their use; hence, a proper understanding of the impact of AAR on the distinct mechanical properties of affected concrete as a function of its induced development is required (Sanchez et al., 2017).

Literature shows that the stiffness (i.e., ME) and tensile strength are the most influenced mechanical properties of AAR-deteriorated concrete (Sanchez et al., 2017). The ME is primarily governed by the mechanical properties of the aggregates, especially the coarse aggregate (Mehta & Monteiro, 2017). Therefore, the deterioration caused by AAR in the aggregate particles is responsible for the significant decrease in ME of affected

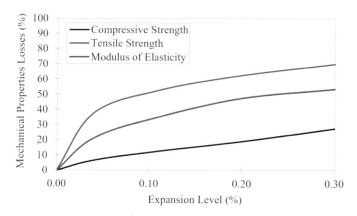

Figure 2.5 Overall mechanical properties (compressive strength, tensile strength and modulus of elasticity) losses of ASR-affected concrete specimens.

concrete. Results demonstrate that losses in ME may vary from 30% at low expansion levels to 65% at high levels of expansion. Likewise, the tensile strength of the concrete is much more affected by AAR than its compressive strength (Giaccio et al., 2008). It has been verified that tensile strength reductions of AAR-affected concrete may drop by about 70%, even at low or moderate levels of expansion (Sanchez et al., 2017). On the other hand, compressive and direct shear strengths seem to be less affected, although some contradictory results are often found, especially in shear (Souza et al., 2019). Numerous experimental campaigns verified that compressive strength is a mechanical property that is not considerably affected by AAR, at least in its early stages. It has been reported that AAR-affected structures displaying expansion levels lower than 0.10% may, in general, still efficiently withstand service loads. Moreover, significant reductions in compressive strength are only observed for high and very high expansion levels (i.e., > 0.20%) as per (Sanchez et al., 2017, 2018; Kubo & Nakata, 2012). Figure 2.5 illustrates the average values obtained by Sanchez et al. (2017) of compressive strength, tensile strength and ME reductions as a function of AAR-induced expansion.

2.3 DELAYED ETTRINGITE FORMATION (DEF)

DEF is a form of ISA, and it is defined as the formation of ettringite in concrete after its setting (or at least a substantial portion of the whole setting) without the contribution of any supplementary external sulphate source from the surrounding environment (Taylor et al., 2001). Besides being considered as a type of ISA (described in Section 2.4.2), DEF can be distinguished due to the "nature" of ettringite formation. Usually, DEF takes place when the concrete undergoes temperatures > 65°C in the very early hours after pouring. Cement hydration and calcium silicate hydrate (C-S-H) formation are greatly accelerated with the increase in the curing temperature. With sustained temperatures above 65°C, ettringite becomes thermodynamically unstable; therefore, previously formed ettringite decomposes, and the sulphate ions return to the solution (Taylor et al., 2001), being absorbed by C-S-H. Later, when SO_4^{2-} is desorbed, the reformation of ettringite leads to induced expansion and cracking. Figure 2.6 illustrates the appearance of DEF-affected columns after 15 years of service.

Although ettringite is commonly seen as homogeneously distributed into the cement paste, DEF-induced expansion and damage in concrete are quite heterogeneous. The reaction kinetics (i.e., induction period, expansion rate and ultimate expansion) depends upon several parameters, such as temperature, sulphate concentration of the cement, environmental conditions (i.e., temperature and humidity) and alkali-loading of the concrete.

2.3.1 Induced expansion mechanism

To better understand the unique damage mechanism of DEF, a brief discussion on the hydration of the aluminate phases of the clinker (C_3A and C_4AF), especially C_3A, is required. The reaction of C_3A with water is immediate; crystalline hydrates, such as C_3AH_6, C_4AH_{19} and C_2AH_8, are quickly formed, releasing a large amount of heat. Gypsum is commonly added to the anhydrate clinker to control such rapid hydration. In general, gypsum and alkalis are quickly released into the concrete pore solution moments after the contact between anhydrate cement and water. Hence, once calcium and aluminum ions become available, there is the precipitation of calcium

(a)   (b)

Figure 2.6 Typical appearance of columns affected by DEF after approximately 15 years (Thomas et al., 2008).

aluminate trisulphate hydrate (AFt – ettringite). Ettringite is usually the first hydrate to crystallize due to the high sulphate/aluminum ratio in the aqueous phase during the first hour of hydration. Later, after most of the sulphate ions are consumed in the Aft's formation, the aluminate concentration rises again, and ettringite becomes unstable, being converted into calcium aluminate monosulphate hydrate (AFm), the final product of Portland cement's hydration bearing over 5% of C_3A.

DEF driven by curing temperatures above 65°C changes the concentration of sulphate and aluminum ions in the concrete pore solution, thus modifying the overall hydration process of the aluminate phases of the clinker. Such high temperatures are normally achieved in steam-cured concrete members or massive structures due to their significant temperature rise during cement hydration. In these conditions, sulphate and aluminate ions tend to be entrapped into C-S-H. Thus, once concrete temperature drops to ambient temperature at later stages, most of the aluminum ions remain firmly bound into C-S-H, whereas SO_4^{2-} is released in the concrete pore solution and continues the overall ettringite formation. Since the concentration of the aluminum ions in the solution is low, most of the ettringite is, therefore, likely to be formed close to sources Al, such as AFm particles (Taylor et al., 2001). This process develops important crystallization pressure in confined spaces. Furthermore, since AFm is homogeneously distributed within the cement paste, the formation of further ettringite leads to fairly homogeneous swelling of the cement paste, developing an extensive crack formation in the affected concrete.

2.3.2 DEF microscopic damage features in concrete

The most acceptable DEF mechanism is characterized by a fairly homogenous expansion of the cement paste (in the presence of moisture), forming circular or peripheral cracks (i.e., gaps, bands) around the aggregate particles, "detaching" them from the cement paste. Thus, "gaps" at the ITZ are progressively generated and filled with large amounts of ettringite (Martin, 2010). Moreover, DEF-induced swelling provides hydraulic tension inside the aggregate particles and radial (in tension), and tangential (in compression) stresses in the cement paste. The radial tension presents its peak at the ITZ; thus, cracks are often developed at these locations. Figure 2.7 presents the common damage features generated by DEF in concrete. If the tensile strength of the aggregate particle is lower than the tensile strength of the ITZ, the fracture surface occurs at locations just inside the aggregate particle (outlining the particle's shape; Goltermann, 1995).

According to Sanchez et al. (2020) DEF-affected concrete displays already at low and moderate expansion levels (i.e., $\leq 0.12\%$) an important number of open cracks in the cement paste. At this level, although most of the cracks are observed in the paste, some cracks can also be verified within the aggregate particles, which might be due to the important stress concentration

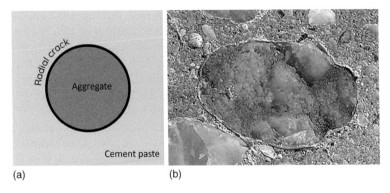

(a)   (b)

Figure 2.7 Chemical/physical induced damage features (or aggregate shrinkage): (a) scheme with a single aggregate particle and (b) affected concrete specimen (Golterman, 1995).

taking place at the ITZ during DEF-induced development. As the expansion increases to high levels (i.e., up to 0.30%), the cracks previously generated keep increasing (in length and width), but new cracks are also formed in the system. Furthermore, some cracks are observed within the aggregate particles. At very high expansion levels (i.e., $\geq 0.30\%$), a major crack network is observed in the cement paste, leading to an important physical integrity loss of the affected material. For expansion levels greater than 0.50%, two new phenomena are observed: debonding and disaggregation of the aggregate particles, which increase, even more, the overall damage of the material.

Recent studies suggest the presence of different damage mechanisms in concrete, such as FT, AAR (as displayed in Figure 2.8), wetting/drying cycles

Figure 2.8 C-beam specimen showing signs of DEF deterioration at Texas A&M Riverside campus (Karthik et al., 2016).

or even the introduction of mechanical loads, may increase the likelihood of DEF-induced development. Moreover, the combination of DEF and ASR is chemically self-supported, and thus, this combined mechanism is often observed in the field. It has been found that ASR reaction products enable the mobility of the ions and thus facilitate DEF-induced development. Furthermore, numerous authors believe that DEF and ASR coupling occurs because the consumption of alkalis due to ASR development to form ASR-secondary products contributes towards DEF development (Thomas et al., 2008; Martin, 2010). Finally, although quantifying the actual contribution of each mechanism towards the overall deterioration process of the concrete is not an easy task, it is commonly agreed that DEF plays a major role in the combined deterioration process (Thomas et al., 2008).

2.3.3 DEF influence on mechanical properties of affected concrete

Evaluating DEF-induced expansion and microscopic damage features discussed in Section 2.3.2 helps the understanding of its impact on the mechanical properties of the affected concrete. According to Sanchez et al. (2018), at expansion levels up to 0.12%, DEF-affected concrete may display a ME 50% lower than sound concrete. Likewise, the tensile strength of the concrete may drop about 70%, even at low or moderate levels of expansion, while the compressive strength reduction may lessen by 10%. The explanation for the low reduction in compressive strength is that (1) the cracks are highly localized at the ITZ and (2) upon loading, existing cracks are partially stopped as per the so-called arrest mechanism provided by the aggregates in the system (Mindess et al., 2003). As the expansion level rises (i.e., up to 0.30%), the ME decreases even more, but at a lower rate (down to about 60%). On the other hand, the compressive strength reduction significantly increases, reaching values of about 40%. At this stage, it is possible to note a trend where some of the ITZ cracks formed at distinct locations start linking to one another, which creates an important net cracking at the cement paste and decreases the physical integrity of the affected material. For expansion levels higher than 0.50%, two new phenomena start taking place: debonding and disaggregation of the aggregate particles, which accelerate once more the drop in stiffness (i.e., ME) of the affected concrete down to values of about 85%. Likewise, the compressive strength drop reaches values of about 50% at this stage. Yet, the cracking extension process at this level of expansion does not seem to further affect the tensile strength of the concrete since these cracks have already reached their "critical length" to cause tensile failure at low to moderate levels of expansion. Figure 2.9 illustrates the average compressive strength, tensile strength and ME losses as a function of DEF-induced expansion of affected concrete (Sanchez et al., 2018; Yammine et al., 2020; Karthik et al., 2016).

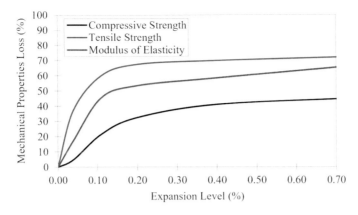

Figure 2.9 Overall mechanical properties (compressive strength, tensile strength and modulus of elasticity) losses of DEF-affected concrete specimens.

2.4 OTHER MECHANISMS

2.4.1 Freezing and thawing cycles

It is well established that porous materials containing moisture are susceptible to deterioration under repeated FT. Freezing water has proven to be quite deleterious to concrete, yet several factors can influence the FT mechanism. Amongst them, the main factors are (a) aggregate features (i.e., particle size, porosity and permeability), (b) the presence of entrained air, (c) the water-to-cement ratio and (d) the degree of saturation of the concrete. Hydraulic pressure is generated when water freezes in the concrete capillarity pores, causing induced expansion since frozen water swells at about 9% of its original volume. Thus, repeatable FT cycles may cause severe physical damage in concrete, such as internal cracking and surface scaling. Figure 2.10 illustrates a concrete sidewalk deteriorated by FT damage.

2.4.1.1 Induced expansion mechanism

When ice forms in the capillary pore structure of concrete, it can lead to induced expansion and microcracking mainly caused by (a) hydraulic pressure generation due to ice formation, (b) osmotic pressure generation due to solute concentration increase in the pore solution adjacent to freezing sites, (c) water desorption from C-S-H and (d) ice segregation. The most common processes presented in the literature are the first two, hydraulic (caused by the increase in the specific volume of water through freezing in large cavities) and osmotic pressure (due to salt concentration differences in the pore water/solution; Mindess et al., 2003). It is generally believed that osmotic pressure predominantly damages concrete exposed to saltwater (e.g., from

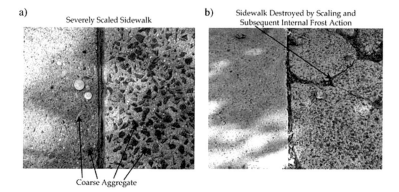

Figure 2.10 (a) Picture of a severely scaled sidewalk. (b) Sidewalk that was severely scaled, which led to a complete loss of mechanical integrity, probably by internal frost damage (Valenza & Scherer, 2006).

de-icing salts, seawater), while hydraulic pressure governs frost deterioration whether no external saltwater is present (Fagerlund, 1994).

Water in the capillary pores of concrete is not pure; it contains several soluble substances, such as alkalis, chlorides and calcium hydroxide. Moreover, it is quite common to find different salt concentrations in the bulk concrete, causing gradient osmotic pressures (Mehta & Monteiro, 2017). Likewise, it is observed that the exposed surface of concrete members cools faster and freezes first. As freezing occurs in the first surface layer of the concrete, water can no longer move through the frozen region. Therefore, continued freezing accompanied by expansion forces water away from the freezing front into the concrete. The result is the pressure exceeding the tensile strength of the cement paste, causing microcracking. This hypothesis also explains why FT-induced deterioration in concrete often occurs with the presence of cracks parallel to the exposed surface (Powers, 1945; Deschenes Jr., 2017).

As previously mentioned, several factors may influence FT-induced mechanisms. For instance, the aggregates' features (i.e., size, porosity and permeability) used in concrete are essential in the damage process. According to Verbeck and Landgren (1996), the aggregates used in concrete can be divided into three categories: (a) first-class aggregates (i.e., low permeability), which are normally not harmed by FT cycles; (b) second-class aggregates (i.e., average permeability), where their particle size is critical since the larger the particle size, the greater the risk of FT damage; and (c) third-class aggregates (i.e., with high permeability), which enable easy water penetration, inducing FT damage in the ITZ rather than within their particles.

Air-entraining admixtures are widely known as FT inhibitors; normally, air-entraining admixtures are selected as a volume percentage (%) of the concrete. However, the most important concept behind the use of air entrainment is not the total volume (%) of entrained air but rather the incorporation of a given amount of round empty voids (which act as valves) spaced between 0.1 and 0.2 mm from one another in the hardened cement paste. However, for practical reasons, the entrained air in (%) is the parameter typically required while mix-proportioning concrete mixtures. Values ranging from 3% to 7% of air, depending on the maximum aggregate size and the surrounding environment, are often selected for concrete mixtures in cold regions. These values are considered very effective in proportion to risk-free FT mixtures, especially if aggregates which are not FT susceptible are used.

In general, when the concrete water-to-cement ratio increases, the mechanical properties of the material decrease due to the rise in overall porosity (i.e., mainly capillary pores). Since water freezes primarily in the concrete pores, the higher the concrete porosity, the higher its likely susceptibility to FT deterioration. However, higher mechanical properties do not necessarily lead to proper FT performance since the presence of air entrainment agents is crucial. Therefore, air-entrained conventional concrete mixtures are expected to behave better against FT-induced deterioration than non-air-entrained, high-performance concrete (Mehta & Monteiro, 2017).

2.4.1.2 FT microscopic damage features in concrete

Following Sanchez et al. (2020) findings, it has been observed that at the beginning of FT-induced development and for low expansion levels (i.e., 0.05%), cracks are mainly formed in the ITZ or the bulk cement paste/pores. Moreover, at this expansion level, a few cracks may also be observed within the aggregate particles. As the expansion level increases and for moderate expansion levels (i.e., 0.12%), the cracks previously formed at the beginning of the deterioration process increase in length and width, and some new cracks are also generated in the cement paste (i.e., ITZ or bulk-paste/pores) and aggregate particles. For high expansion levels (i.e., 0.20%), the distress mechanism keeps progressing, either through the increase of pre-existent cracks or by forming new cracks in the cement paste. Otherwise, the generation of cracks within the aggregate particles seems to stop completely. Finally, when the expansion level reaches very high degrees (i.e., 0.30% and beyond), most of the cracks present in the cement paste link to one another, resulting in an important crack network formation.

2.4.1.3 FT influence on mechanical properties of affected concrete

Evaluating FT-induced expansion and microscopic damage features helps to understand their associated impact on the mechanical properties of the affected concrete. According to Sanchez et al. (2018), for low expansion levels, a significant amount of cracks are found within the aggregate particles, yet with much less intensity than for ASR-affected specimens, for instance. Therefore, the FT-affected specimens may experience a decrease of 35% in ME. Comparably to ASR and DEF, FT-affected concrete displays tensile strength reductions down to 70% at low expansion amplitudes. The failure mechanism in tension is indeed a direct and brittle mechanism dictated by "fracture mechanics", where "stress concentrations" are generated in the presence of micro defects/pores of concrete due to FT-induced development. Likewise, since FT cracks are mainly formed within the cement paste, a much more significant drop in compressive strength, reaching values between 15% and 30% of loss, is observed at low expansions. At moderate expansion levels (i.e., 0.12%), the ME keeps dropping down to values of about 40%, while compressive strength reductions reach values between 20% and 35%, and the tensile strength keeps dropping, yet at a lower rate. As mentioned in Section 2.4.1.2, from expansion amplitudes of 0.12% up to 0.20%, the damage generation in the cement paste keeps progressing, but the cracks within the aggregate particles stop being generated. This distress feature continues at about 0.30% of expansion; thus, the drop in the modulus reaches values of about 50%. The tensile strength loss at this phase seems similar to that obtained at moderate expansion levels, which means that since the tensile strength is extremely dependent on the crack length (i.e., maximum length needed to cause the crack's propagation leading to failure), its threshold was likely already attained at lower expansion levels. In addition, the compressive strength drop reaches about 40%, being more significant than ASR and almost the same as DEF for the same expansion levels. Figure 2.11 illustrates the average compressive strength, tensile strength and ME losses in the function of the expansion amplitudes of FT-affected concrete (Sanchez et al., 2018).

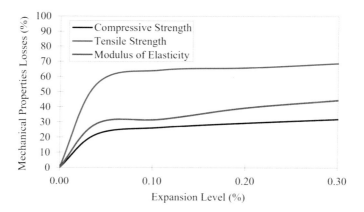

Figure 2.11   Overall mechanical properties (compressive strength, tensile strength and modulus of elasticity) losses of FT-affected concrete specimens.

2.4.2 Internal sulphate attack (ISA)

The term *sulphate attack* comprises complex processes of chemical reactions between sulphate ions (SO_4^{2-}) and hydrated products from the cement paste in the presence of high moisture. Overall, these processes include salt crystallization (i.e., *physical attack*) and chemical attack by sulphate ions from internal (i.e., contaminated materials, sulphate-bearing aggregates or binder materials) or external (i.e., soil, groundwater, seawater, etc.) sources in the concrete. Nevertheless, besides a few particularities of the different named mechanisms of sulphate attack, the global chemistry involving SO_4^{2-} anions in these processes are quite similar; its swelling behaviour is commonly attributed to the formation of gypsum and secondary ettringite. Equation 2.3 illustrates the formation of gypsum from the reaction of sulphate ions and portlandite and the formation of secondary ettringite from its further reaction with calcium monosulfoaluminate hydrate (AFm):

$$SO_4^{2-} + Ca(OH)_2 + 2H_2O \rightarrow CaSO_4.2H_2O + 2OH^-$$

$$\qquad \text{Portlandite} \qquad\qquad\qquad\qquad \text{Gypsum} \qquad\qquad\qquad\qquad\qquad (2.3)$$

$$2CaSO_4.4H_2O + 4CaO.Al_2O_3.SO_3.12H_2O$$

$$\text{Gypsum} \qquad\qquad\qquad\qquad \text{AFm}$$

$$+ 16H_2O \rightarrow 6CaO.Al_2O_3.3SO_3.32H_2O$$

$$\qquad\qquad\qquad \text{Ettringite} \qquad\qquad\qquad\qquad\qquad (2.4)$$

It is worth noting that ettringite is a common product that originates from cement hydration; its formation only becomes harmful when it occurs after the concrete's setting. The formation of gypsum and secondary ettringite commonly yields changes in volume by about 1.2 to 2.2 times higher than the reactant products, leading to an increase in the internal pressure on its surroundings and causing volumetric deformation, cracking, spalling and mechanical properties reductions. Moreover, the magnitude of the induced damage development depends upon several parameters, such as sulphate concentration, humidity, exposure and curing temperature, along with the cation associated with SO_4^{2-}.

Most standards worldwide suggest restrictive limits on both sulphur trioxide and C_3A contents of cement; the purpose of this double restriction is to prevent internal/external sulphate attack. Such restriction was recognized when it was observed that extensive induced expansion might occur in the absence of an external source of sulphate. By definition, ISA takes place when there is a delayed release of sulphates in the hardened cement paste or in the presence of iron sulphide-bearing aggregates (e.g., aggregates bearing pyrrhotite, pyrite). In general, pyrrhotite and pyrite minerals are oxidized (as shown in Equations 2.5 and 2.6, respectively) in the presence of water and oxygen, producing ferrous ions (Fe^{2+}) and sulphate (SO_4^{2-}), and releasing hydrogen ions (H^+) into the cement paste. Besides the further reaction of the SO_4^{2-} to form gypsum and secondary ettringite, the oxidation of pyrrhotite/pyrite leads to a decrease in the pH of the concrete pore solution, accelerating the oxidation of the minerals (Dobrovolski et al., 2021). Moreover, the oxidation rate of sulphide minerals increases directly with increasing relative humidity to values up to 80% (Steger, 1982).

$$Fe_{1-x}S + \left(2 - x/2\right)O_2 + xH_2O \rightarrow \left(1 - x\right)Fe^{2+} + SO_4^{2-} + 2xH^+ \tag{2.5}$$

$$FeS_2 + 7/2O_2 + H_2O \rightarrow Fe^{2+} + 2SO_4^{2-} + 2H^+ \tag{2.6}$$

Concrete deterioration due to the combined effects of the oxidation of iron sulphides (e.g., in sulphide-bearing aggregates as seen in Figure 2.12a) and ISA in the cement paste (i.e., due to the release of sulphide ions) generates secondary minerals with high swelling properties, causing expansion development in both concrete microstructure phases, aggregates and cement paste. The expansion in the aggregate particles is commonly attributed to the oxidation of iron, similar to steel corrosion in reinforced concrete. Petrographic evaluations, illustrated in Figure 2.12b, demonstrate that important cracks are observed around and through the aggregate particles towards the cement paste of affected concrete (Rodrigues et al., 2012). Moreover, the aggregate particles are completely covered with iron oxyhydroxide and are surrounded by a whitish halo; the latter was verified as being a region rich in ettringite.

Furthermore, changing humidity and temperature can either intensify/accelerate the damage development or even change the mechanism of ISA.

(a)   (b)

Figure 2.12 Features of concrete deterioration caused by iron sulphide-bearing aggregates: (a) Cracking in a house's foundations. Open cracks are typically more pronounced at the corners of the foundation blocks, often next to rain gutters. Yellowish surface coloration is often seen on the exposed foundation walls. (b) Stereomicroscopic views of a deteriorated concrete foundation block (Rodrigues et al., 2012).

For instance, high-temperature conditions during the hardening process of the concrete can lead to further development of *DEF* (described in Section 2.3), whereas low temperature conditions can lead to *TF*, whose mechanism will be further discussed in the following section.

2.4.3 Thaumasite formation (TF)

In recent years, sulphate attack caused by TF has received considerable attention. TF involves sulphate and carbonate ions in the presence of high humidity and "preferentially" low temperature (< 15°C). However, TF has already been reported at higher temperatures (Bassuoni & Nehdi, 2009). Differently from the pyrrhotite/pyrite oxidation or DEF, TF mainly "targets" the SO_4^{2-} ions that are present in the C-S-Hs (the main binding phase in concrete) rather than the ones present in the calcium hydroxide and calcium aluminate phases. Equation 2.7 demonstrates thaumasite chemical formation. In severe cases of TF, besides the intense expansion development proportioned by the sulphate attack and ettringite formation, the hardened cement paste is expressively replaced by thaumasite, transforming the concrete into a white and incohesive mush (Skalny et al., 2002).

$$3CaO \cdot 2SiO_2 \cdot 3H_2O + 2Ca(OH)_2 + 2SO_4^{2-} + 2CaCO_3$$
$$\underset{\text{C–S–H}}{} \qquad \underset{\text{Portlandite}}{}$$
$$+ 24H_2O \rightarrow 2(CaSiO_3 \cdot CaCO_3 \cdot CaSO_4 \cdot 15H_2O)$$
$$\underset{\text{Thaumasite}}{} \tag{2.7}$$

It is worth mentioning that TF may occur in any type of sulphate attack as long as carbonate ions and preferably low temperatures are included in the process. Moreover, it is unlikely that thaumasite occurs without the formation of ettringite. Most case studies of TF-related deterioration show that thaumasite and ettringite coexist in the microstructure of concrete (Rahman & Bassuoni, 2014). According to the literature, there are two possible "routes" for TF: direct and indirect. In the former, sulphate ions react with carbonate (either CO_3^{2-} ions or atmospheric CO_2) and C-S-H to form thaumasite, as demonstrated in Equation 2.7, whereas in the indirect "route," ettringite is a precursor for TF (Bensted, 2003). In other words, the damage development taking place in TF-affected concrete tends first to follow the primary source of SO_4^{2-} release; however, with the worsening factor of expressive strength loss due to the deterioration of C-S-H. Finally, it is worth mentioning that the mechanism of TF is still under debate in the scientific community; therefore, an in-depth investigation of TF conditions is still required and could shed some light on preventive measures against TF-induced development.

2.4.4 Physical sulphate attack (PSA)

The so-called *PSA*, also called *sulphate salt crystallization* or *salt hydration distress*, usually refers to repeated cycles of recrystallization of *mirabilite* ($Na_2SO_4 \cdot 10H_2O$) into *thenardite* (Na_2SO_4), which lays on the constant variation of both humidity and temperature. This temperature-dependent process leads to repeated volume variation, illustrated by Figure 2.13, leading to fatigue of the cement paste and its subsequent loss of cohesion (Skalny et al., 2002). First, high moisture conditions solubilize sodium sulphate salts (internally or externally to concrete), which facilitates its transport and diffusion through concrete. Afterwards, once the exposure conditions change and the water starts evaporating, the sodium and sulphate ions "reattach" to one another, forming sodium sulphate decahydrate ($Na_2SO_4 \cdot 10H_2O$). Moreover, the continuous changes in temperature or humidity over time are followed by the sodium sulphate anhydrite (Na_2SO_4) and vice versa.

In a PSA, the involvement of the hydrated Portland cement products is not "mandatory". However, the damage mechanism of PSA and whether it is independent of chemical sulphate attack is still controversial. The reaction and crystal formation commonly take place in any empty flaws within the concrete, such as pre-existent cracks and further developed cracks, air voids, pores in the cement paste or aggregates, at the ITZ. Scaling and flaking of the concrete's surface are typical signs of PSA because salts tend to crystallize in pores around the evaporation surfaces. Figure 2.14 illustrates deterioration due to PSA.

2.4.5 Hydration of crystalline MgO and CaO

The expansive effect of crystalline magnesium oxide hydration in concrete was recognized more than a hundred years ago (i.e., 1884) when the first cases were reported in France and Germany, while it was quite common to find cement containing about 25% of MgO in its chemical composition

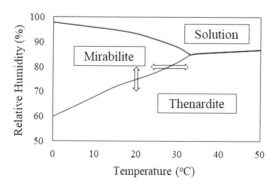

Figure 2.13 Temperature-dependent phase diagram for sodium sulphate (Tsui et al., 2003).

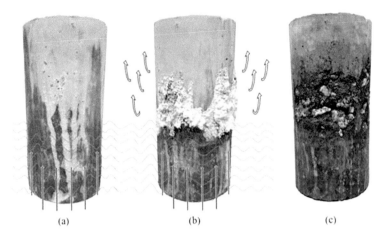

Figure 2.14 Mechanism of PSA damage with (a) specimen placed in partial immersion, (b) salts wick up through the concrete and deposit on its surface and (c) resulting surface damage (Alyami et al., 2019).

(Mehta & Monteiro, 2017). Nowadays, different standards (e.g., ASTM C150, 2018) require that the MgO content in cement shall not exceed 6%. However, MgO expansive agents are commonly used to compensate for temperature shrinkage of mass concrete (Mo et al., 2014). Although approximately 118% of solid volume increase is caused when MgO transforms into $Mg(OH)_2$, the corresponding induced expansion is not precisely equal to the increase of solid volume. Overall, MgO hydrates slowly, and it is associated with delayed induced expansion in concrete through the formation of $Mg(OH)_2$ (i.e., brucite). There are several theories developed to explain the expansion mechanism of MgO (Mo et al., 2014); the two main common theories are (a) crystal growth pressure due to the growth of $Mg(OH)_2$ and (b) swelling pressure due to water absorption by very small crystals of $Mg(OH)_2$. During the hydration of MgO, a supersaturated solution of Mg^{2+} and OH^- are firstly formed; then $Mg(OH)_2$ crystals precipitate and grow in a confined region, resulting in crystal growth pressure leading to induced expansion (Chatterji, 1995). Nevertheless, the most important factors affecting the expansion of MgO concrete are the dosage and reactivity of MgO and temperature (i.e., the higher the curing temperature, the higher the hydration kinetics of MgO).

The harmful effect due to the hydration of partially crystalline CaO (poorly burnt from $CaCO_3$) was reported in the 1930s in the United States. Although the presence of a high volume of partially crystalline CaO in cement can cause a severe induced expansion in concrete, this phenomenon is virtually unseen in modern concrete construction due to the better manufacturing protocols of a Portland cement clinker. However, the driving force leading to induced expansion while the hydration of partially crystalline CaO is the crystal growth pressure generated after later $Ca(OH)_2$ formation under restricted conditions near the surfaces of poorly burnt lime grains (Chatterji, 1995). Therefore, the solid volume of $Ca(OH)_2$ formed (from the later hydration of CaO) doubles the initial volume of CaO and thus, expansive forces are generated, causing the swelling of the cement paste and further cracking. Yet, the severity of the induced expansion may be decreased in low alkali systems, likely due to the release of calcium ions in the concrete pore solution (Min et al., 1996).

REFERENCES

Alyami, M. H., Alrashidi, R. S., Mosavi, H., Almarshoud, M. A., & Riding, K. A. (2019). Potential accelerated test methods for physical sulphate attack on concrete. *Construction and Building Materials 229*, 116920. https://doi.org/10.1016/j.conbuildmat.2019.116920

ASTM C150. 2018). Standard Specification for Portland Cement, issued 2018.

Bassuoni, M. T., & Nehdi, M. L. (2009). Durability of self-consolidating concrete to different exposure regimes of sodium sulphate attack. *Materials and Structures/Materiaux et Constructions 42*, 1039–1057. https://doi.org/10.1617/s11527-008-9442-2

Bensted, J. (2003). Thaumasite – direct, woodfordite and other possible formation routes. *Cement and Concrete Composites 25*(8), 873–77. https://doi.org/10.1016/S0958-9465(03)00115-X

Bérard, J., & Roux, R. (1986). La Viabilité Des Bétons Du Québec: Le Rôle Des Granulats. *Canadian Journal of Civil Engineering 13*(1), 12–24. https://doi.org/10.1139/l86-003

Chatterji, S. (1995). Mechanism of expansion of concrete due to the presence of dead-burnt CaO and MgO. *Cement and Concrete Research 25*(1), 51–56. https://doi.org/10.1016/0008-8846(94)00111-B

Deschenes Jr. R. A. (2017). Mitigation and Evaluation of Alkali-Silica Reaction (ASR) and Freezing and Thawing in Concrete Transportation Structures. ProQuest Dissertations and Theses. Ann Arbor: University of Arkansas. https://login.proxy.bib.uottawa.ca/login?url=https://www.proquest.com/dissertations-theses/mitigation-evaluation-alkali-silica-reaction-asr/docview/1949666139/se-2?accountid=14701

Dobrovolski, M. E. G., Munhoz, G. S., Pereira, E., & Medeiros-Junior, R. A. (2021). Effect of crystalline admixture and polypropylene microfiber on the internal sulphate attack in portland cement composites due to pyrite oxidation. *Construction and Building Materials 308*(September), 125018. https://doi.org/10.1016/j.conbuildmat.2021.125018

Dunant, C. F., & Scrivener, K. L. (2010). Micro-mechanical modelling of alkali-silica-reaction-induced degradation using the AMIE framework. *Cement and Concrete Research 40*(4), 517–25. https://doi.org/10.1016/j.cemconres.2009.07.024

Fagerlund, G. (1994. The Water Absorption Process in the Air-Pore System WATER ABSORPTION PROCESS IN THE AIR-PORE SYSTEM Report TVBM-7085.

Fournier, B., & Bérubé, M.-A. (2000). Alkali-aggregate reaction in concrete: A review of basic concepts and engineering implications. *Canadian Journal of Civil Engineering 27*(2), 167–91. https://doi.org/10.1139/cjce-27-2-167

Giaccio, G., Zerbino, R., Ponce, J. M., & Batic, O. R. (2008). Mechanical behavior of concretes damaged by alkali-silica reaction. *Cement and Concrete Research 38*(7), 993–1004. https://doi.org/10.1016/j.cemconres.2008.02.009

Goltermann, P. (1995). Mechanical predictions of concrete deterioration; Part 2: Classification of crack patterns. *ACI Materials Journal 92*(1), 1–6. https://doi.org/10.14359/1177

Karthik, M. M., Mander, J. B., & Hurlebaus, S. (2016). Deterioration data of a large-scale reinforced concrete specimen with severe ASR/DEF deterioration. *Construction and Building Materials 124*, 20–30. https://doi.org/10.1016/j.conbuildmat.2016.07.072

Katayama, T. (2010). The So-Called Alkali-Carbonate Reaction (ACR) - Its mineralogical and geochemical details, with special reference to ASR. *Cement and Concrete Research 40*(4), 643–75. https://doi.org/10.1016/j.cemconres.2009.09.020

Katayama, T., & Grattan-Bellew, P. E. (2012). Petrography of the Kingston Experimental Sidewalk at Age 22 Years–ASR as the Cause of Deleteriously Expansive, So-Called Alkali-Carbonate Reaction. In *Proceedings of the 14th International Conference on Alkali-Aggregate Reaction in Concrete*. Austin, Texas, 10.

Kubo, Y., & Nakata, M. (2012). Effect of reactive aggregate on mechanical properties of concrete affected by alkali-silica reaction. In *14th International Conference on Alkali-Aggregate Reaction in Concrete*, electronic. Austin (Texas).

Leemann, A., & Lura, P. (2013). E-Modulus of the alkali-silica-reaction product determined by micro-indentation. *Construction and Building Materials 44*, 221–27. https://doi.org/10.1016/j.conbuildmat.2013.03.018

Martin, R. P. (2010). Analyse Sur Structures Modèles Des Effets Mécaniques de La Réaction Sulfatique Interne Du Béton. Université Paris-Est.

Mehta, P. K., & Monteiro, P. J. M. (2017. "CONCRETE Microstructure, Properties and Materials."

Min, D., Dongwen, H., Xianghui, L., & Mingshu, T. (1996). Mechanism of expansion in hardened cement pastes with hard-burnt lime. *Cement and Concrete Research 26*(4), 647–48. https://doi.org/10.1016/0008-8846(96)00021-x

Mindess, S., Young, J. F., & Darwin, D. (2003). *Concrete*. Prentice-Hall Civil Engineering and Engineering Mechanics Series. Prentice-Hall. https://books.google.ca/books?id=38VoQgAACAAJ

Mo, L., Deng, M., Tang, M., & Al-Tabbaa, A. (2014). MgO expansive cement and concrete in China: Past, present and future. *Cement and Concrete Research 57*, 1–12. https://doi.org/10.1016/j.cemconres.2013.12.007

Powers, T. C. (1945). A working hypothesis for further studies of frost resistance of concrete. *ACI Journal Proceedings 41*(1). https://doi.org/10.14359/8684

Rahman, M. M., & Bassuoni, M. T. (2014). Thaumasite sulphate attack on concrete: Mechanisms, influential factors and mitigation. *Construction and Building Materials 73*(12), 652–62. https://doi.org/10.1016/j.conbuildmat.2014.09.034

Rajabipour, F., Giannini, E., Dunant, C., Ideker, J. H., & Thomas, M. D. A. (2015). Alkali-Silica reaction: Current understanding of the reaction mechanisms and the knowledge gaps. *Cement and Concrete Research 76*(6), 130–46. https://doi.org/10.1016/j.cemconres.2015.05.024

Reinhardt, H. W., & Mielich, O. (2011). A fracture mechanics approach to the crack formation in alkali-sensitive grains. *Cement and Concrete Research*. https://doi.org/10.1016/j.cemconres.2010.11.008

Rodrigue, A., Duchesne, J., Fournier, B., Champagne, M., & Bissonnette, B. (2020). Alkali-Silica reaction in Alkali-activated combined slag and fly ash concretes: The tempering effect of fly ash on expansion and cracking. *Construction and Building Materials 251*. https://doi.org/10.1016/j.conbuildmat.2020.118968

Rodrigues, A., Duchesne, J., Fournier, B., Durand, B., Rivard, P., & Shehata, M. (2012). Mineralogical and chemical assessment of concrete damaged by the oxidation of sulfide-bearing aggregates: Importance of thaumasite formation on reaction mechanisms. *Cement and Concrete Research 42*(10), 1336–1347. https://doi.org/10.1016/j.cemconres.2012.06.008

Sanchez, L. F.M., Drimalas, T., Fournier, B., Mitchell, D., & Bastien, J. (2018). Comprehensive damage assessment in concrete affected by different Internal Swelling Reaction (ISR) mechanisms. *Cement and Concrete Research 107* (February), 284–303. https://doi.org/10.1016/j.cemconres.2018.02.017

Sanchez, L. F. M., Drimalas, T., & Fournier, B. (2020). Assessing condition of concrete affected by Internal Swelling Reactions (ISR) through the Damage Rating Index (DRI). *Cement 1–2*(September), 100001. https://doi.org/10.1016/j.cement.2020.100001

Sanchez, L. F. M., Fournier, B., Jolin, M., & Duchesne, J. (2015). Reliable quantification of AAR damage through assessment of the Damage Rating Index (DRI). *Cement and Concrete Research 67*(1), 74–92. https://doi.org/10.1016/j.cemconres.2014.08.002

Sanchez, L. F. M., Fournier, B., Jolin, M., Mitchell, D., & Bastien, J. (2017). Overall assessment of Alkali-Aggregate Reaction (AAR) in concretes presenting different strengths and incorporating a wide range of reactive aggregate types and natures. *Cement and Concrete Research 93*, 17–31. https://doi.org/10.1016/j.cemconres.2016.12.001

Sanchez, L. F. M. (2014). Contribution to the assessment of damage in aging concrete infrastructures affected by alkali-aggregate reaction. *341*.

Skalny, J., Marchand, J., & Odler, I. (2002). *Sulphate attack on concrete* (1st ed.). London and New York: Spon Press.

Souza, D. J. De, Sanchez, L. F. M., & De Grazia, M. T. (2019). Evaluation of a direct shear test setup to quantify AAR-induced expansion and damage in concrete. *Construction and Building Materials 229*, 116806. https://doi.org/10.1016/j.conbuildmat.2019.116806

Steger, H. F. (1982). Oxidation of sulfide minerals. VII. Effect of temperature and relative humidity on the oxidation of pyrrhotite. *Chemical Geology 35*(3–4), 281–295. https://doi.org/10.1016/0009-2541(82)90006-7

Taylor, H. F. W., Famy, C., & Scrivener, K. L. (2001). Delayed ettringite formation. *Cement and Concrete Research 31*(5), 683–693. https://doi.org/10.1016/S0008-8846(01)00466-5

Thomas, M. D. A., & Folliard, K. J. (2007). Concrete aggregates and the durability of concrete. *Durability of Concrete and Cement Composites*, 247–281. https://doi.org/10.1533/9781845693398.247

Thomas, M., Folliard, K., Drimalas, T., & Ramlochan, T. (2008). Diagnosing delayed ettringite formation in concrete structures. *Cement and Concrete Research 38*(6), 841–847. https://doi.org/10.1016/j.cemconres.2008.01.003

Tsui, N., Flatt, R. J., & Scherer, G. W. (2003). Crystallization damage by sodium sulphate. *Journal of Cultural Heritage 4*(2), 109–115. https://doi.org/10.1016/S1296-2074(03)00022-0

Valenza, J. J., & Scherer, G. W. (2006). Mechanism for salt scaling. *Journal of the American Ceramic Society 89*(4), 1161–1179. https://doi.org/10.1111/j.1551-2916.2006.00913.x

Verbeck, G., & Landgren, R. (1996). Influence of physical characteristics of aggregates on frost resistance of concrete. In *Proceedings of the American Society for Testing and Materials (ASTM '96)*, 1063–1079. Philadelphia, USA.

Yammine, A., Leklou, N., Choinska, M., Bignonnet, F., & Mechling, J. M. (2020). DEF damage in heat cured mortars made of recycled concrete sand aggregate. *Construction and Building Materials 252*, 119059. https://doi.org/10.1016/j.conbuildmat.2020.119059

Chapter 3

Assessing the condition of ISR-affected concrete

3.1 SERVICE LIFE OF CONCRETE INFRASTRUCTURE

Critical concrete infrastructure, including bridges, overpasses, dams, tunnels, industrial buildings and stadiums, is designed with a specified lifespan known as service life. The concept of service life dates back to early observations made by builders who noticed that certain materials and approaches had longer lifespans than others (ACI 365.1R-00 2000; Davey 1961). Historically, the prediction and evaluation of service life have relied on qualitative and empirical methods. However, advancements in understanding the common deterioration mechanisms of concrete have paved the way for more quantitative predictions of concrete infrastructure lifespan (ACI 365.1R-00 2000).

Typically, critical infrastructure is designed with a service life ranging from 50 to 75 years, depending on the structure type and design standards. The service life of such infrastructure is not solely determined by the risk of failure but also considers overall functionality. Excessive operating costs, for example, may render a structure economically unviable, which in turn may require its replacement and, thus, the end of service life. It is important to note that the terms "durability" and "service life" are distinct and are often mistakenly used interchangeably. Durability refers to the ability of a product, component, assembly or construction to maintain its intended function(s) over a specified time, while service life is defined as the period after placement during which all properties consistently exceed the minimum acceptable values with routine maintenance (ACI 365.1R-00 2000).

According to the ACI 365 committee, three types of service life can be defined: technical service life, functional service life and economic service life (Sommerville 1986; Sommeville 1992):

- *Technical service life*: elapsed time until a defined unacceptable state is reached, such as significant concrete spalling, safety levels falling below acceptable thresholds or member's failure.

DOI: 10.1201/9781003188155-3

- *Functional service life*: refers to the duration until a structure no longer meets functional requirements or becomes obsolete due to changing demands, such as the need for increased clearance or road widening.
- *Economic service life*: the duration at which replacing the structure or a portion of it becomes more financially advantageous than maintaining it in service.

Service life concepts and methodologies are applicable in both the design and operation phases of structures. Parameters such as water-to-cementitious materials ratio, concrete cover and inspection and maintenance strategies are considered in the design phase to ensure the desired service life. Over the years, various methodologies have been developed to predict the service life of critical concrete infrastructure. These methodologies rely on information about the current condition of the structure (or structural member) under analysis, deterioration rates, past and future loading conditions and definition of the end of service life (Clifton 1991). Based on predictions of remaining life, economic decisions can be made regarding whether to repair, rehabilitate or replace the structure. The end of service life is typically determined by criteria such as (ACI 365.1R-00 2000)

- severe material deterioration or exceeding the design load-carrying capacity,
- maintenance requirements surpass available resources,
- aesthetics becoming unacceptable and
- functional capacity of the structure no longer meeting demand.

Ultimately, decisions regarding the end of service life are influenced by considerations of human safety and economics. In many cases, the condition, appearance or capacity of concrete structures can be improved to acceptable levels; however, the associated costs may be prohibitive. Therefore, management strategies for critical concrete infrastructure should be based upon a thorough assessment of the structure's condition (ACI 365.1R-00 2000).

3.2 CONDITION ASSESSMENT OF CONCRETE INFRASTRUCTURE

Condition assessment refers to the comprehensive evaluation of the overall state of a structure or its components. Traditional protocols for condition assessment begin with a detailed examination of relevant information pertaining to the structure, including soil profiles, foundation type, as-built drawings, construction records and a history of deterioration. This initial phase is followed by field surveys and testing that encompass visual inspections, the application of non-destructive testing (NDT), structural health monitoring (SHM) and long-term monitoring techniques. In certain cases, laboratory and structural analyses may also be employed as necessary (Blight & Alexander 2011; Omar & Nehdi 2018; Saviotti 2014). Once the aforementioned tasks have been completed, structural evaluations are conducted to assess the current load-bearing capacity of the structure (or structural member(s)) under analysis and ascertain its ability to withstand service loads. These evaluations involve the use of structural testing and/or modelling techniques (ACI 365.1R-00 2000; Blight & Alexander 2011; Kennedy 1958). Figure 3.1 illustrates the common sequence of activities while the condition assessment of concrete infrastructure.

A comprehensive and systematic condition assessment is essential for accurately predicting the current and future performance of infrastructure, as well as for optimizing maintenance, rehabilitation and replacement strategies. However, it is worth noting that many ageing structures are still evaluated using simplistic, qualitative and subjective methods such as visual inspection, which can be influenced by the personal judgement of the inspector. Nonetheless, significant advancements have been made in recent decades in the field of advanced techniques for condition assessment, including NDT, SHM, laboratory analyses and more. In the following sections, a brief overview of these techniques will be provided, discussing their fundamental concepts and specific applications, both individually and in combination, for achieving a reliable assessment of critical infrastructure.

3.2.1 Examination of current documentation

Prior to conducting any on-site inspection, it is essential to thoroughly review all relevant documents and information pertaining to the structure. This comprehensive review serves as the foundation for the subsequent field inspection. The following information should be carefully sought and examined:

- Identification of the structure type and purpose
- Assessment of environmental conditions and the degree of exposure
- Date of completion and any historical records
- Documentation of prior rehabilitation or maintenance work, whether minor or major
- Details regarding the design loading and construction specifics
- Examination of foundation characteristics and soil profiles
- Evaluation of the drainage system and joint arrangements
- Identification of the reinforcement, including its positioning, amount and detailing
- Documentation of concrete properties, such as mix-proportions and the sources of raw materials employed (e.g., cement and aggregates)
- Review of previous inspection reports and any past field and laboratory analyses conducted

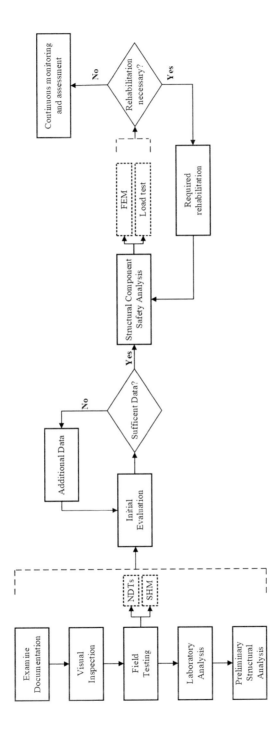

Figure 3.1 Condition assessment flowchart.

(Adapted from ACI 365.1R-00 2000; Omar & Nehdi 2018.)

- Identification of the dates when signs of deterioration were first observed
- Consideration of any additional information that may be relevant to the field inspection, such as the performance of neighbouring or similar structures, age of the structure and any design modifications that have been made

Once all the necessary information has been gathered, field inspections can be scheduled. It is important to note that the validity and reliability of the available documents, particularly test results, should be carefully evaluated and, if possible, cross-verified through on-site observations.

During the site inspection, visible indications of deterioration must be meticulously documented. This includes recording crack widths, directions, spacing and precise locations. These observations should be captured photographically and annotated on scale drawings or, alternatively, recorded through freehand sketches with measured dimensions and directional indicators, such as a north point (Blight & Alexander 2011; Fournier et al. 2010).

3.2.2 Visual inspection (VI)

VI serves as the primary task in conducting a comprehensive condition assessment of concrete infrastructure. In various regions, including North America, routine VI is typically performed at intervals of approximately two years, with the specific frequency dependent on the type and condition of the structure being analysed. For ageing infrastructure (i.e., over 30 years old), enhanced VI is conducted at six-year intervals. Additionally, detailed emergency inspections are performed when there is an imminent risk of failure or following catastrophic failure to ensure public safety (Akula et al. 2014; Omar & Nehdi 2018). VI plays a crucial role in detecting and identifying signs and patterns of deterioration, including flaws, defects, crack networks, discolorations and the presence of secondary products associated with various deterioration mechanisms. However, it is important to note that VI is a qualitative and subjective procedure influenced by the equipment used and operator interpretation. In many cases, it may not provide an accurate assessment of the cause and extent of deterioration or facilitate the selection of appropriate rehabilitation strategies for affected structures (Blight & Alexander 2011; Moore et al. 2000; Zahedi et al. 2022). Various equipment can be utilized during VI to aid in the inspection process, such as (Blight & Alexander 2011; Fournier et al. 2010)

- hand lens,
- binocular,
- crack gauge,
- measuring tape and
- camera.

However, there is a pressing need for the development of improved and standardized VI guidelines tailored to different types of structures. This would ensure that VI outcomes can be reliably utilized in the overall assessment of critical infrastructure, particularly those affected by ISR. Chapter 4 explores both conventional and advanced VI approaches employed for assessing critical concrete infrastructure.

3.2.3 Field testing

Several techniques can be employed in the field to assess the condition of critical infrastructure, with three of the most commonly used being NDT, SHM and load testing (LT).

3.2.3.1 Non-destructive testing (NDT)

The early detection of deterioration in concrete, prior to major damage, is facilitated by the use of NDT. NDT methods can be integrated into VI protocols for concrete infrastructure to evaluate parameters such as stiffness, strength, moisture content and the presence of flaws, defects and cracks. Over the past decades, several NDT methods have been developed, each exploring distinct phenomena, such as acoustic, seismic, electric, thermal and electromagnetic, based on the predominant deterioration mechanisms leading to damage (Gucunski et al. 2013). For instance, the probability of active corrosion can be assessed using methods including half-cell potential (Pradhan & Bhattacharjee 2009), electrical resistivity (Browne 1980) and ground penetrating (Varnavina et al. 2015) techniques, while corrosion rate can be identified by the polarization resistance method (Cady & Gannon 1992). Furthermore, the presence of vertical cracks, which results in a reduced modulus of elasticity of concrete, can be captured using the ultrasonic surface wave method (Nazarian et al. 1993). Concrete delamination can be detected using techniques such as impact echo (Kee et al. 2012; Shokouhi et al. 2014), pulse echo (Krause et al. 2011) and infrared thermography tests (Kee et al. 2012; Washer et al. 2009). Additional information on the theoretical foundations, instrumentations, applications and data analysis of NDT can be found in Chapter 4.

3.2.3.2 Structural health monitoring (SHM)

SHM is a non-destructive technique employed for in situ evaluation. It utilizes multiple sensors embedded in the structural member(s) of interest to monitor their structural response and identify any abnormal behaviour. The primary objective of SHM techniques is to estimate deterioration and assess its impact on the structural response and capacity. In recent years, several SHM systems have been developed and implemented to provide valuable information on concrete infrastructure. These systems typically share fundamental elements, including sensor measurements and instrumentation, structural assessment through peak strains or modal analysis and support for maintenance, rehabilitation and replacement strategies (Alampalli 2012). The effectiveness of an SHM system depends on the type and quantity of sensors employed. Monitoring systems can utilize single or multiple sensor types, which can be customized to capture various physical measurements associated with loads and environmental conditions (Wong 2007). In addition to structural evaluation, SHM systems, equipped with diverse sensor types, can also identify material properties such as concrete creep, shrinkage and corrosion. Furthermore, they can assess environmental effects, such as temperature gradients and dynamic responses (e.g., traffic-induced vibrations; Feng & Feng 2018). While SHM methods hold great promise, their widespread implementation is still limited. The absence of technical standards for SHM systems has historically hindered their extensive adoption in the condition assessment of critical infrastructure (Bulletin fib Task Group 3.3. 2022; TG3.3 fib Bulletin 2023).

3.2.3.3 Load testing (LT)

LT is a valuable technique for assessing critical infrastructure, especially bridges (AASHTO 2011). It enables the determination of safe loading levels for structures or structural members. Through forced static and dynamic load testing in varied load conditions, the maximum response can be detected using strain transducers placed at critical locations of the structure or structural member of interest (Omar & Nehdi 2018). Load tests can be broadly classified into two categories: proving load tests, which serve as self-supporting alternatives to theoretical assessments, and supplementary load tests, which complement theoretical calculations (Zhang et al. 2016). Load ratings are then established using methods such as allowable stress, load factor or load and resistance factor. Additionally, load testing can be combined with preliminary or advanced structural analysis, including finite element models (FEM), as well as NDT and SHM techniques. This integration, within a structural reliability framework, facilitates the determination of practical safety factors for ageing concrete infrastructure (Wang et al. 2011).

3.2.3.4 Field testing combination for assessing the condition of concrete infrastructure

Various condition assessment protocols and frameworks have been developed over the years based on the field methodologies discussed earlier. One of the most recent and comprehensive approaches was introduced by the International Federation for Structural Concrete (TG3.3 fib Bulletin 2023). This framework revolves around the acquisition and utilization of a set of indicators that pertain to the serviceability and safety of the structure (or structural member) being analysed. These indicators are derived from data obtained through different yet complementary methods, including VI, NDT and SHM. By utilizing these methodologies, the framework enables the calculation of indicators that provide insight into the current state and loading conditions of the structure. The indicators are classified based on the method used to gather the information and the level of assessment, whether it is local or global. It is important to highlight that within this proposed approach, NDT is particularly suited for localized damage detection evaluation, while SHM is typically utilized to obtain global performance data for the structure or its components (Figure 3.2) (Bulletin fib Task Group 3.3., 2022).

3.2.4 Laboratory analyses

The condition of the structure (or structural member) under analysis may require additional laboratory tests. These tests encompass various techniques, such as chemical, microscopic and mechanical assessments, each serving its own purpose. Chemical assessment focuses on investigating the presence of secondary products that can indicate or confirm the occurrence of chemical deterioration in concrete. This deterioration can lead to material degradation and, consequently, structural implications. Microscopic analyses are indicated to evaluate concrete microstructure, which could suggest the presence of secondary products and or crack patterns related to physical or chemical deterioration. Mechanical assessment searches on the impact of physical or chemical deterioration on the engineering properties of concrete, such as tensile, compressive and shear strengths, as well as stiffness. These methods are typically selected when there is a significant degree of visible damage to the structure under investigation or when ISR mechanisms are suspected to be involved in the overall damage process. More detailed discussions on these procedures can be found in Chapters 5 (microscopic techniques) and 6 (mechanical testing).

3.2.5 Structural analyses

A thorough structural analysis should be conducted using either traditional or numerical methods to assess the structural performance of concrete infrastructure after field and/or laboratory testing. Traditional methods rely on conventional design standards and guidelines, offering effective approaches but often characterized by conservative assumptions. On the other hand, numerical assessment, such as FEM, enables a more reliable, accurate and realistic representation of the structure's condition (Omar & Nehdi 2018). In FEM, various variables are typically considered, including the properties of the materials, the geometry of the structure and/or structural components, the design and detailing of reinforcement, the current state of damage, the construction process and the environmental conditions

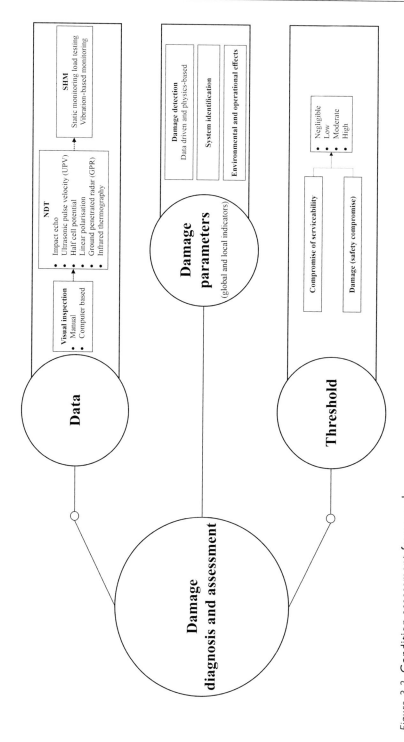

Figure 3.2 Condition assessment framework.

(Adapted from Bulletin fib Task Group 3.3., 2022.)

(Sousa et al. 2014). Therefore, the use of FEM is extremely popular to evaluate the structural capacity of ISR-affected infrastructure.

3.3 ISRS AS ONGOING DETERIORATION PROCESSES

Significant progress has been made over the years in the field of condition assessment of concrete infrastructure. Traditional assessment methods, including some of the methodologies previously discussed in this chapter, have proven effective for evaluating ageing and deterioration processes, such as load-induced cracks or damage caused by aggressive external agents (e.g., carbonation, chloride penetration leading to steel corrosion, sulphate attack), where the deterioration rates caused by these mechanisms are fairly well understood via transport mechanisms, besides being easily detected (or measured) in the field. However, when internal damage mechanisms, such as ISRs, occur within the concrete material, the application of these tools and protocols becomes more complex and requires further examination.

ISR mechanisms are harmful, ongoing processes leading to induced expansion and deterioration which impact the durability, serviceability and structural performance of affected structures. The rate of induced expansion and deterioration varies depending on factors such as the type of ISR (e.g., alkali-aggregate reaction, delayed ettringite formation, freeze and thaw (FT) damage), environmental conditions (global and local factors like temperature, relative humidity, exposure degree and solar radiation), concrete mix-proportions and the geometry and confinement of the affected structures (or structural members). It is generally accepted that ISR-induced expansion follows an S-shaped curve over time, although linear trends are often observed in practice, especially in large structures such as dams. Figure 3.3 provides an illustration of the theoretical evolution of ISR-induced expansion as a function of time.

The assessment of structures affected by ISR is a complex undertaking since besides detecting *the main cause(s)* generating damage, along with their current *deterioration extent*, the so-called *diagnosis*, it is also important to forecast their *future behaviour over time* (i.e., *prognosis*). The techniques mentioned earlier in this chapter, although useful, offer only a limited assessment of ISR-affected structures. They may provide insights into either the cause(s) or the extent of the damage but generally yield short-term information that is insufficient for predicting the long-term performance and service life of the affected structures (Kennedy 1958). This means that the development of a comprehensive and reliable protocol enabling the evaluation of the current cause(s) and degree of damage, along with the deterioration rate over time (e.g., coupling laboratory test data with in situ monitoring and modelling) of ISR-affected concrete, is crucial and yet a quite complex task, requiring engineering analysis, judgement and interpretation (ACI 365.1R-00 2000).

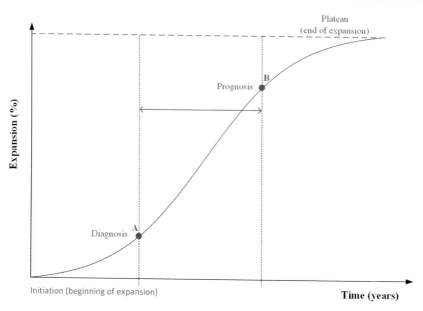

Figure 3.3 Theoretical evolution of ISR-induced expansion over time.

The proposed protocol should integrate service history, materials and structural members' features, current damage cause(s) and extent, structural analyses and a comprehensive degradation model to enable proper prediction of future behaviour. Furthermore, such methodology should be able to appraise the role of maintenance in extending service life or structural reliability of damaged structures (Naus & Oland 1994; Sanchez et al. 2020). Moreover, it is essential to recognize that different ISR mechanisms, such as AAR, FT and DEF, exhibit unique deterioration processes that have varying impacts on the engineering properties of the affected concrete. Therefore, it becomes necessary to define the concept of "damage" in a broader and more comprehensive manner that encompasses these distinctions. While the term "damage" has been interpreted differently in various studies and by different authors, it can be broadly defined as the "*harmful and measurable consequences of various mechanisms (e.g., loadings, shrinkage, creep, ASR, DEF, freezing and thawing) on the mechanical properties, physical integrity, and durability of concrete*" (Sanchez et al. 2020). In practical terms, damage can be defined as (a) engineering properties reductions (i.e., compressive, tensile and direct shear strengths) of concrete, (b) stiffness reduction of concrete and (c) physical integrity and/or durability loss, which is directly related to the existing cracking network. Figure 3.4 illustrates the broad damage definition for ISR-affected concrete (Sanchez et al. 2020).

It is important to note that it is outside of the scope of the book to discuss available techniques, models and calculations to be used to perform structural assessment of ISR-affected structures. However, the primary objective

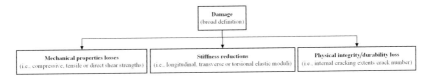

Figure 3.4 Global classification of damage in concrete.
(Adapted from Sanchez et al. 2014.)

of this book is to comprehensively examine the efficiency of various techniques and protocols that can be employed to assess the current and future condition of ISR-affected concrete. This is crucial to provide input and data for subsequent structural performance analyses. In this context, various visual, non-destructive, microscopic and mechanical techniques have been utilized to assess the condition of ISR-affected concrete. Some of these techniques have shown considerable promise, while others have been deemed obsolete. In the forthcoming chapters, the implementation, adaptation and advancement of field and laboratory techniques capable of efficiently quantifying ISR-induced deterioration will be extensively discussed. The advantages and disadvantages of each technique will be presented, and protocols enabling the development of frameworks to support infrastructure owners in making informed decisions on their assets will be introduced.

REFERENCES

AASHTO. (2011). *Manual for Bridge Evaluation* (2nd ed.). American Association of State Highway and Transportation Officials (AASHTO). ISBN: 978-1-56051-519-7
ACI 365.1R-00. (2000). *Service-Life Prediction-State-of-the-Art Report.*
Akula, M., Zhang, Y., Kamat, V. R., & Lynch, J. P. (2014). Leveraging structural health monitoring for bridge condition assessment. *Construction Research Congress 2014*, 1159–1168. https://doi.org/10.1061/9780784413517.119
Alampalli, S. (2012). Special Issue on Nondestructive evaluation and testing for bridge inspection and evaluation. *Journal of Bridge Engineering*, 17(6), 827–828. https://doi.org/10.1061/(ASCE)BE.1943-5592.0000430
Blight, G. E., & Alexander, M. G. (2011). *Alkali-aggregate reaction and structural damage to concrete*. CRC Press. https://doi.org/10.1201/b10773
Browne, R. D. (1980). Mechanisms of corrosion of steel in concrete in relation to design, inspection, and repair of offshore and coastal structures. *ACI Special Publication, 65*, 169–204.
Bulletin fib Task Group 3.3. (2022). *Existing concrete structures: Life management, testing and structural health monitoring (in preparation)*.
Cady, P. D., & Gannon, E. J. (1992). *Condition evaluation of concrete bridges relative to reinforcement corrosion. Volume 8: Procedure manual*. National Research Council.
Clifton, J. R. (1991). *Predicting the remaining service life of concrete and concrete technology*. National Institute of Standards and Technology.

Davey, N. (1961). *A history of building materials*. Phoenix House.

Feng, D., & Feng, M. Q. (2018). Computer vision for SHM of civil infrastructure: From dynamic response measurement to damage detection – A review. *Engineering Structures*, *156*, 105–117. https://doi.org/10.1016/j.engstruct.2017.11.018

Fournier, B., Bérubé, M. A., Folliard, K., & Thomas, M. (2010). *Report on the diagnosis, prognosis, and mitigation of Alkali-Silica Reaction (ASR) in transportation structures*.

Gucunski, N., Imani, A., Romero, F., Nazarian, S., Yuan, D., Wiggenhauser, H., Shokouhi, P., Taffe, A., & Kutrubes, D. (2013, January 13). Nondestructive testing to identify concrete bridge deck deterioration. *Proceedings of the 92nd Meet*. https://doi.org/10.17226/22771

Kee, S.-H., Oh, T., Popovics, J. S., Arndt, R. W., & Zhu, J. (2012). Nondestructive bridge deck testing with air-coupled impact-echo and infrared thermography. *Journal of Bridge Engineering*, *17*(6), 928–939. https://doi.org/10.1061/(ASCE) BE.1943-5592.0000350

Kennedy, T. B. (1958). Laboratory testing and the durability of concrete. In ASTM STP 236 (Ed.), *Symposium on Approaches to Durability in Structures*. ASTM.

Krause, M., Mayer, K., Friese, M., Milmann, B., Mielentz, F., & Ballier, G. (2011). Progress in ultrasonic tendon duct imaging. *European Journal of Environmental and Civil Engineering*, *15*(4), 461–485. https://doi.org/10.1080/19648189.2011. 9693341

Moore, M., Phares, B., Graybeal, B., Rolander, D., & Washer, G. (2000). *Reliability of visual inspection for highway bridges*. Federal Highway Administration (FHWA).

Naus, D. J., & Oland, C. B. (1994). *Structural aging program technical progress report for period*.

Nazarian, S., Baker, R., & Crain, K. (1993). *Development and testing of a seismic pavement analyzer; report SHRP-H-375*.

Omar, T., & Nehdi, M. (2018). Condition assessment of reinforced concrete bridges: Current practice and research challenges. *Infrastructures*, *3*(3), 36. https://doi. org/10.3390/infrastructures3030036

Pradhan, B., & Bhattacharjee, B. (2009). Half-cell potential as an indicator of chloride-induced rebar corrosion initiation in RC. *Journal of Materials in Civil Engineering*, *21*(10), 543–552. https://doi.org/10.1061/(ASCE)0899-1561(2009)21:10(543)

Sanchez, L. F. M., Fournier, B., Jolin, M., & Bastien, J. (2014). Evaluation of the stiffness damage test (SDT) as a tool for assessing damage in concrete due to ASR: Test loading and output responses for concretes incorporating fine or coarse reactive aggregates. *Cement and Concrete Research*, *56*, 213–229. https://doi. org/10.1016/j.cemconres.2013.11.003

Sanchez, L. F. M., Fournier, B., Mitchell, D., & Bastien, J. (2020). Condition assessment of an ASR-affected overpass after nearly 50 years in service. *Construction and Building Materials*, *236*, 117554. https://doi.org/10.1016/j. conbuildmat.2019.117554

Saviotti, A. (2014). Bridge assessment, management and life cycle analysis. *Modern Applied Science*, *8*(3). https://doi.org/10.5539/mas.v8n3p167

Shokouhi, P., Wolf, J., & Wiggenhauser, H. (2014). Detection of delamination in concrete bridge decks by joint amplitude and phase analysis of ultrasonic array measurements. *Journal of Bridge Engineering*, *19*(3). https://doi.org/10.1061/(ASCE) BE.1943-5592.0000513

Sommerville, G. (1986). Design life of structures. *The Structural Engineer, 64A*(2), 60–71.

Sommeville, G. (1992). Service life prediction – an overview. *Concrete International, 14*(11), 45–49.

Sousa, H., Bento, J., & Figueiras, J. (2014). Assessment and management of concrete bridges supported by monitoring data-based finite-element modeling. *Journal of Bridge Engineering, 19*(6). https://doi.org/10.1061/(ASCE)BE.1943-5592.0000604

TG3.3 fib Bulletin. (2023). *Condition assessment of reinforced concrete structures: State of the art: knowledge and case studies in the TG3.3 fib Bulletin.*

Varnavina, A. V., Khamzin, A. K., Torgashov, E. V., Sneed, L. H., Goodwin, B. T., & Anderson, N. L. (2015). Data acquisition and processing parameters for concrete bridge deck condition assessment using ground-coupled ground penetrating radar: Some considerations. *Journal of Applied Geophysics, 114*, 123–133. https://doi.org/10.1016/j.jappgeo.2015.01.011

Wang, N., Ellingwood, B. R., & Zureick, A.-H. (2011). Bridge rating using system reliability assessment. II: Improvements to bridge rating practices. *Journal of Bridge Engineering, 16*(6), 863–871. https://doi.org/10.1061/(ASCE)BE.1943-5592.0000171

Washer, G., Fenwick, R., Bolleni, N., & Harper, J. (2009). Effects of environmental variables on infrared imaging of subsurface features of concrete bridges. *Transportation Research Record: Journal of the Transportation Research Board, 2108*(1), 107–114. https://doi.org/10.3141/2108-12

Wong, K.-Y. (2007). Design of a structural health monitoring system for long-span bridges. *Structure and Infrastructure Engineering, 3*(2), 169–185. https://doi.org/10.1080/15732470600591117

Zahedi, A., L. F. Sanchez, & Noël, M. (2022). Appraisal of visual inspection techniques to understand and describe ASR-induced development under distinct confinement conditions. *Construction and Building Materials, 323*, 126549. https://doi.org/10.1016/j.conbuildmat.2022.126549

Zhang, Q., Alam, M. S., Khan, S., & Jiang, J. (2016). Seismic performance comparison between force-based and performance-based design as per Canadian Highway Bridge Design Code (CHBDC) 2014. *Canadian Journal of Civil Engineering, 43*(8), 741–748. https://doi.org/10.1139/cjce-2015-0419

Chapter 4

Visual inspection and non-destructive testing (NDT)

4.1 INTRODUCTION

Visual inspection (VI) and non-destructive testing (NDT) are frequently the first procedures conducted while assessing the condition of concrete infrastructure, regardless of the damage type and degree observed. These techniques allow for gathering preliminary information on the presence of deterioration while deciding if further assessment is required (Kim et al. 2019). This chapter presents a thorough description of visual inspection procedures along with commonly used NDT to assess the condition of ISR-affected concrete.

4.2 VISUAL INSPECTION (VI)

VI is normally the very first step towards assessing the presence of deterioration, irregularities and displacements of concrete members, ensuring they still meet structural safety regulations and expected service requirements (Henrickson et al. 2016; Koch et al. 2015). VI processes generally include a descriptive survey of the type(s) and degree of existing deterioration observed on the surface of concrete members (David & Gregory 2017; Gattulli & Chiaramonte 2005; Kabir 2010a). Although VI is qualitative by nature, it can be combined with quantitative measurements, such as the measurement of crack widths and computation of crack density, to provide more detailed information and classification on the overall deterioration process. For instance, Table 4.1 illustrates the accepted crack widths in reinforced concrete members under service loads as per the American Concrete Institute (ACI) 224R-01 – Control of cracking in concrete structures (ACI 224R-19 2019).

Regardless of the mechanism inducing damage, performing routine VI is considered crucial to identify early deterioration and thus prevent damage escalation and important structural implications leading to catastrophic failure (Golden et al. 2018; Koch et al. 2015; Wood 2008). Moreover,

DOI: 10.1201/9781003188155-4

Table 4.1 Acceptable crack widths in reinforced concrete members under service loads (ACI 224R-19 2019)

	Crack width	
Exposure condition	*mm*	*inches*
Dry air or protective membrane	0.41	0.016
Humidity, moist air, soil	0.30	0.012
De-icing chemicals	0.18	0.007
Seawater and seawater spray, wetting and drying	0.15	0.006
Water-retaining structures (excluding nonpressure pipes)	0.10	0.004

although the full recognition of the main cause(s) leading to deterioration cannot be made only through VI, the knowledge of common damage patterns and features associated with distinct deterioration processes may help in their identification in the field. The next section presents common "macroscopic" damage features related to ISR in affected concrete.

4.2.1 ISR-induced surface damage signs on concrete

4.2.1.1 Alkali-aggregate reaction (AAR)

The main damage indicators and features associated with AAR-induced deterioration are as follows (ACI 201.1R 2008):

- *Deposits*: upon water penetration in concrete, deposits may be generated on the concrete member's surface due to the dissolved or leached chemicals. Figure 4.1 illustrates distinct deposits that may be presented as (a) efflorescence, where whitish and powdery salt is observed; (b) exudation, where a liquid or gel-like deposit is observed in voids or cracks; (c) incrustation, where a hard crust is formed and; (d) stalactite, where a downward pointing formation suspended from the concrete surface is formed, usually shaped like an icicle.
- *Map cracking*: A typical indication of AAR-induced development where cracks form a repetitive polygonal pattern that partially or locally covers the surface of affected concrete. These randomly distributed cracks may become more oriented in locations bearing higher confinement and/or restraint degrees. Figure 4.2 illustrates a y-shaped column seated in a foundation block. The block displays a random map cracking since it bears minor reinforcement, whereas the cracks present in the y-shaped columns follow a more oriented path (i.e., parallel to the main reinforcement) due to the reinforcement restraint.

(a)   (b)

(c)   (d)

Figure 4.1 Deposit signs on the surface of AAR-deteriorated concrete: (a) efflorescence along major cracks and joints, (b) exudation of gel-like product staining the concrete adjacent to the cracks, (c) encrustation on the surface of the concrete and (d) stalactites.

([a] photograph courtesy of Leandro Sanchez; [b–d] photograph courtesy of Cassandra Trottier.)

Figure 4.2 Alkali-silica reaction (ASR)-affected concrete member in Quebec City, Canada.

(Photograph courtesy of Leandro Sanchez.)

(a)   (b)

Figure 4.3 Typical DEF damage features in massive concrete members from (a) (Thomas et al. 2008) and (b) (Godart & Divet 2013).

4.2.1.2 Delayed ettringite formation (DEF)

The main damage indicators and features associated with DEF are similar to AAR (Martin et al. 2015). The main difference is that DEF-induced expansion and deterioration typically occur in massive concrete structures such as piers, pier caps, bridge girders and foundation blocks, as illustrated in Figure 4.3; DEF is also observed in steam-cured concrete, normally used in precast members (Godart & Divet 2013; Karthik et al. 2016b; Karthik et al. 2016a).

4.2.1.3 Other mechanisms: freezing and thawing (FT)

Deterioration caused by fluctuating temperatures is largely observed under harsh climates where the mean temperature varies above and below the freezing point of water (i.e., 0°C); this variation leads to the occurrence of FT cycles, which may cause a wide range of deterioration signs, including the following (ACI 201.1R 2008):

- *Scaling*: Local flaking or peeling of finished surfaces of hardened concrete. Scaling can be categorized as marginal, moderate and severe depending on the deterioration depth (d) and area, i.e., marginal scaling (loss of surface mortar, no coarse aggregates' exposure); moderate scaling (loss of surface mortar from 5 < d < 10 mm and some aggregate's exposure) as displayed in Figure 4.4a; severe scaling (loss of surface mortar from 5 < d < 10 mm surrounding 10 or 20 mm coarse aggregate particles; very severe (loss of surface mortar and coarse aggregate particles, with d > 20 mm), as per Figure 4.4b.
- *Delamination*: A type of scaling affecting a larger area, where two layers of concrete are separated from one another, as illustrated by Figure 4.4c; delamination may not appear at the surface of affected concrete but is detectable through NDT.

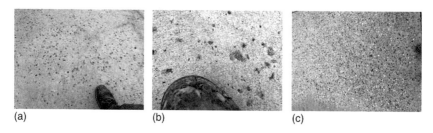

(a)   (b)   (c)

Figure 4.4 Concrete surface scaling at two levels of damage: (a) small scaling showing small dispersed aggregate particles, (b) medium scaling showing dispersed aggregate particles and some pits of mortar and (c) severe scaling showing a large number of exposed aggregate particles.

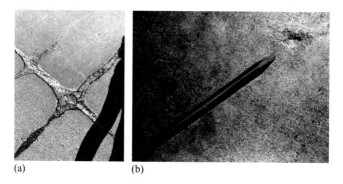

(a)   (b)

Figure 4.5 (a) Typical D-cracking (Auberg & Setzer 2014) and (b) pop-out in concrete.

(Photograph courtesy of Cassandra Trottier.)

- *Cracking*: FT-induced cracking is generally propagated in the form of the so-called D-cracking, which is near and roughly parallel to joints and edges. Such damage can be observed in concrete pavements after a few years of service, as illustrated in Figure 4.5a.
- *Pop-outs*: damage feature defined by the spalling of small portions of concrete incorporating coarse aggregate particles, as shown in Figure 4.5b; normally, pop-outs take place during the use of highly permeable coarse aggregate particles that break under pressure caused by water freezing.

4.2.1.4 Other mechanisms: sulphide-bearing aggregates

The deterioration caused by sulphide-bearing aggregates has gained significant attention in the last couple of years in regions such as North America and northern Europe (Casanova et al. 1996; Chinchón et al. 1995; Lugg & Probert 1996), especially after the occurrence of important deterioration in Trois-Rivière (Quebec, Canada), where concrete foundations in a number of houses exhibited quite impressive damage signs within a relatively short period of time (i.e., three to five years; Duchesne et al. 2021; Rodrigues et al. 2012). The most common deterioration signs of sulphide-bearing aggregates deterioration include (Rodrigues et al. 2012):

- *Cracking*: As for AAR and DEF, sulphide-bearing aggregate induces randomly distributed cracks in concrete, the so-called map cracking (Figure 4.6a), with greater prominence at the edges of affected structural members or areas where water accumulates, such as locations close to rain gutters (Figure 4.6b). The crack openings can reach up to 10 mm, and in regions of more severe damage, these values can escalate to as high as 40 mm (Figure 4.6c).
- *Deposits and discolouration*: The surface of concrete affected by sulphide-bearing aggregates often displays a yellowish coloration. Iron hydroxides and rust can be observed within the cracks, indicating

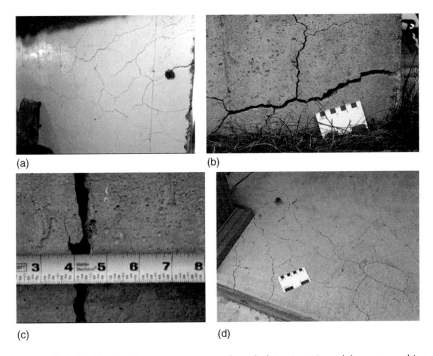

(a)   (b)

(c)   (d)

Figure 4.6 Sulphide-bearing aggregates induced deterioration: (a) map-cracking pattern on interior home foundation painted wall in Connecticut, United States (Geiss & Gourley 2019); (b) cracking observed at a house foundation in Trois-Rivière, Canada; the top green bars of the crack comparator card represents 1 cm and the bottom represents 1″ (Rodrigues et al. 2016); (c) a 12 mm (1/2″) crack opening in a concrete foundation in Connecticut, United States (Zhong & Wille 2018); and (d) a slab-on-grade with yellowish and rust stains in Trois-Rivière, Canada (Rodrigues et al. 2012).

the presence of chemical reactions and oxidation processes. Reddish-brown discolouration (i.e., rust spots or stains) may also be observed ay the concrete's surface, as illustrated in Figure 4.6d (Jana 2022).

- *Pop-out*: Pop-outs are quite common on the interior side of foundation walls. These pop-outs are characterized by the presence of oxidized aggregate particles surrounded by a whitish or yellowish powdery deposit. They commonly occur in areas where the spalling of small portions of concrete, incorporating coarse aggregate particles, has taken place.
- *Excessive deformations*: Due to the extensive concrete cracking and displacement, deformations of concrete members even showing section losses can be observed (Figure 4.7).

4.2.2 Quantifying ISR-induced deterioration in concrete

Semi-quantitative approaches have been developed to complement the outcomes of qualitative and descriptive visual inspections; besides quantifying the surface damage degree of affected concrete, they can be used to estimate the rate of deterioration over time. Amongst the proposed methods, the cracking index (CI) is probably the most conventional technique used to appraise ISR-induced deterioration, providing an indication of the damage degree of the deteriorated concrete member under evaluation.

CI was proposed in France in the 1990s and requires drawing a region of interest (ROI) for analysis, which is normally 0.5 m by 0.5 m square divided into even intervals (Figure 4.8); the cracks intercepting the drawn segments are counted, and their widths measured. Sometimes, a cross is drawn in the interior of the square to increase precision in capturing cracks from distinct directions (Figure 4.8a). The size of the ROI may be adjusted (i.e., increased

Figure 4.7 **A map-cracked concrete house foundation presenting very large crack openings in Connecticut, United States (Rodrigues et al. 2016).**

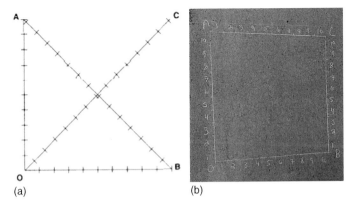

Figure 4.8 CI method: (a) schematic of the original CI as per (Fasseu & Michel 1997) and (b) adapted CI grid drawn on the surface of the deteriorated concrete.

(Photograph courtesy of Cassandra Trottier.)

or decreased) based on the specific structural member being investigated. For larger members such as dams and retaining walls, a square with dimensions of at least 1 m by 1 m is recommended. Otherwise, for slender members, such as precast beams, squares smaller than 0.5 m by 0.5 m are suitable (Figure 4.8b). It is crucial to adapt the ROI to ensure accurate measurements for different types of structural members. Furthermore, CI becomes not only a tool enabling assessment of the current deterioration but also of the deterioration rate once the same ROI is assessed over time (Fasseu & Michel 1997; Fournier et al. 2010; Leemann et al. 2021).

The calculation of the CI involves dividing the total summation of crack openings by the size of the square segments selected (i.e., base length) as illustrated by Equation 4.1:

$$CI = \frac{\sum Crack\ openings}{Base\ length} \tag{4.1}$$

Concrete members presenting *CI* values greater than 0.5 mm/m and/or cracks with widths exceeding 0.15 mm require additional investigations for better identifying the cause(s) leading to deterioration, along with the extent of internal damage (Fasseu & Michel 1997; Fournier et al. 2010; Leemann et al. 2021). Building on the previous equation, Zahedi et al. (2022) proposed a modified version of the CI calculation, allowing the estimation of the attained expansion in reinforced concrete displaying distinct levels of reinforcement, as described in Equation 4.2. Figure 4.9 illustrates the correlation between attained expansion and the proposed approach, allowing a much better estimation when compared solely with the CI value, yet the

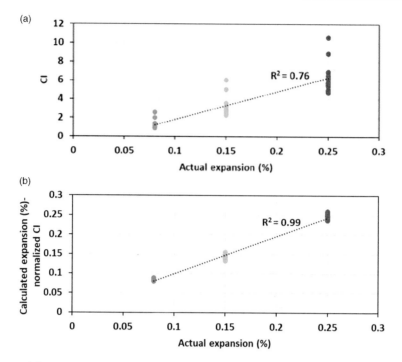

Figure 4.9 Usage of CI outcomes to estimate induced expansion of ISR-affected concrete: (a) correlation between CI and actual expansion (measured via demec points) and (b) correlation between the proposed approach by Zahedi et al. and actual expansion (Zahedi et al. 2022).

application of the proposed approach for concrete members bearing distinct geometry, confinement/restraint and environmental conditions still requires further investigation (Zahedi et al. 2022).

$$\varepsilon\left(\%\right) = \frac{\dfrac{\sum Crack\ openings}{Base\ length}}{n} \tag{4.2}$$

The value of "n" represents the count of cracks observed within the baseline length over which the CI measurement is performed.

4.2.3 Other VI techniques

The rise of new and sophisticated technologies has enabled the development of tools used to perform VI in a more efficient and timely manner, increasing the accessibility of the members under evaluation and, thus, more thorough and frequent inspections. Such works have been applied to a wide range of structures and are normally conducted via image analysis (or digital image processing) techniques. Image analysis is a tool used to quantify objects and patterns in images and can, therefore, complement VI protocols since, amongst others, VI bears the purpose of recognizing distinct damage features and patterns associated with the various types of ISRs. The automatic detection of these numerous damage features has been thus the focus of several studies, from mapping deposits and discoloration (Valença et al. 2013) to identifying potential areas of interest showing signs of corrosion, delamination and efflorescence (Ogawa et al. 2022; Savino & Tondolo 2021; Valença et al. 2013) and cracking in general (Billah et al. 2019; Carrasco et al. 2021; Dias et al. 2021; Jang et al. 2019; Kim et al. 2020; Liu et al. 2020; Ogawa et al. 2022; Savino & Tondolo 2021; Zhu et al. 2011). In the aforementioned works, cameras from smartphones, drones or vehicles were used (Miyamoto 2013; Radopoulou & Brilakis 2017; Zhang et al. 2017, 2018), all of which with the goal to decrease the time and increase the accessibility of visual inspections.

However, only a few works have used image analysis and corresponding automated techniques to visually assess the surface of concrete affected by ISR. Kabir (Kabir 2010b) evaluated the level of damage induced by AAR using a thermographic image coupled with artificial intelligence to characterize and quantify cracking. Moliard (Moliard et al. 2016) digitized images of concrete surfaces affected by ISR and applied an automated version of the French CI while updating the index to detect anisotropic behaviour.

It is expected that automated techniques may significantly enhance the VI and condition assessment of concrete affected by ISR, especially when performed over time (Moliard et al. 2016; Vlahović et al. 2012). Moreover, the progression of the damage and differential movements on the surface of ISR-affected concrete could be obtained by acquiring images at successive time intervals. The analysis of the differential movements, named digital image correlation (DIC), is also amongst the emerging technologies used to monitor concrete structures (Hansen & Hoang 2021; Mahadevan et al. 2017; Thériault et al. 2022). These types of visual inspection techniques are non-destructive in nature and are promising tools to enhance the quality of current VI approaches.

4.3 NON-DESTRUCTIVE TESTING (NDT)

NDT can complement VI since visual damage signs may not appear until reaching moderate- to high-damage levels (Gunn et al. 2017). The concept of NDT is generally based on physical principles such as energy, electricity and motion. NDTs enable the estimation of mechanical properties and the detection of defects, flaws and imperfections, thus facilitating damage assessment with more or less limited accuracy. Despite notable advancements in data processing capabilities and the integration of sophisticated techniques (e.g., reflection attenuation, diffraction, scattering, complex theories of heterogeneous media and digital imaging processing), NDT still possesses limitations in quantitatively evaluating ISR-affected concrete infrastructure. However, these techniques may also provide interesting qualitative or relative appraisal, comparing sound versus damaged locations, for instance.

Various NDTs have been explored for evaluating concrete distressed by ISR in the field. These techniques include electrical resistivity, surface thermography, ground penetrating radar (GPR), stress waves and resonant frequency analysis. The following sections present the most used NDT for assessing critical concrete infrastructure affected by ISR, highlighting their advantages and disadvantages, as well as discussing the future challenges associated with their field implementation.

4.3.1 Electrical resistivity (ER)

ER is an intrinsic property of materials that hinders the flow of electrical current through their bodies, regardless of their geometry. While dry concrete does not exhibit an electrical response, the presence of a pore solution allows for the passage of current. This pore solution contains various ions, including Ca^{++}, K^+, Na^+, OH^- and SO^- (as discussed in Chapter 2), which undergo changes in concentration over time and affect electrical flow. Consequently, ER becomes a sensitive parameter for detecting the progression of ISR.

A range of set-ups can be employed to measure ER, with the Wenner probe and the two-point uniaxial methods being the most widely used for surface and bulk ER measurements, respectively.

The Wenner probe (i.e., surface ER) presents a configuration consisting of four electrodes; these electrodes are evenly spaced and placed on the concrete surface. The outer electrodes apply an alternating current to the concrete, while the two inner electrodes measure the potential (Figure 4.10b).

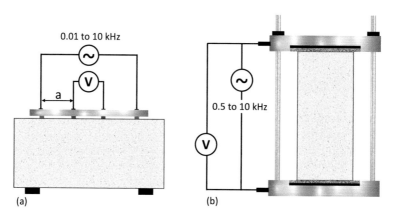

Figure 4.10 Concrete resistivity testing set-ups: (a) surface ER and (b) bulk ER.

By analysing the difference in potential relative to the applied current, the resistivity of the concrete surface is determined. The two-point method (i.e., bulk ER) involves placing the concrete specimen between two parallel metal plates acting as electrodes. It is important to ensure a reliable connection between the specimen and the electrodes, and this can be enhanced by incorporating moist sponges between the metal plates and the specimen. Once the specimen is appropriately positioned and connected to the electrodes, an alternating current is applied, and the resulting voltage or potential reduction, is measured (Figure 4.10a). Detailed information about the testing set-up can be found in (Morris et al. 1996).

While ER initially showed promise for assessing ISR-induced damage, certain challenges have emerged in practice, including the significant influence of moisture conditions on the concrete's ER. Figure 4.11a illustrates the typical trend of ER versus induced expansion, while Figure 4.11b shows the relationship between ER and time. Notably, surface ER demonstrates a sensitive correlation with expansion up to a damage level of 0.05%, beyond which no correlation is observed.

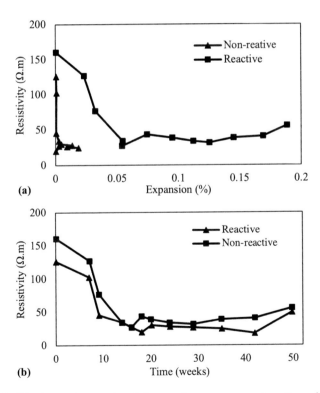

Figure 4.11   ER of non-damaged and damaged concrete as a function of (a) expansion (%) and (b) time. (in weeks).

(Adapted from Rivard & Saint-Pierre 2009.)

Recently, interesting improvements in the sensitivity and accuracy of surface ER for assessing ISR damage have been demonstrated. However, due to variations among results, further clarification is still needed regarding parameters that may directly affect surface ER outcomes. Additionally, although limited tests have been performed, bulk ER has shown interesting correlations with microscopic test procedures used to characterize damage in ISR-affected concrete, such as the damage rating index (DRI) (Grazia 2023; Strow et al. 2022); the DRI will be discussed in detail in Chapter 5.

Besides its application in assessing ISR-damaged concrete, ER has proven useful for evaluating the effectiveness of ISR-preventive measures, such as the use of supplementary cementitious materials (SCMs) in concrete. However, it is important to note that the curing temperature significantly influences the estimation of ISR-induced expansion using surface ER (Chopperla & Ideker 2022; Wang et al. 2022). Finally, given the variety of results obtained, further research is still necessary to better understand and clarify the parameters that directly affect surface ER outcomes when assessing ISR.

4.3.2 Surface thermography (ST)

Surface thermography is a non-intrusive and contactless method that monitors temperature changes on the surface of an object over time using thermographic technologies. By employing an infrared detector, infrared thermography captures thermal patterns on the object's surface, providing valuable insights into both surface and subsurface irregularities (Rogalski 2011). Concrete surfaces can be accurately evaluated via thermal energy using emissivity, allowing for the identification of warmer and cooler regions, as illustrated in Figure 4.12. In the assessment of large concrete areas, the sun serves as a suitable and cost-effective heat energy source, ensuring even heat distribution across the entire surface. It is important to note that surfaces with varying degrees of roughness may absorb radiation at different levels.

Defects in concrete, such as delamination, poor consolidation, water infiltration or ISR deterioration, introduce heterogeneities that result in distinct heat flow patterns when compared to sound regions. ISR-induced cracking in concrete leads to variations in temperature absorption, providing an opportunity to identify damaged regions through the analysis of heat flow across the concrete surface.

Statistical texture analysis is commonly applied, followed by supervised neural network classification approaches to enhance the accuracy of thermography imaging for assessing ISR-affected structures. These techniques enable the extraction of cracking patterns, including measurements of width, length and area (Kabir 2010b).

While infrared thermography shows promise in aiding routine inspection for assessing concrete infrastructure affected by ISR, further research

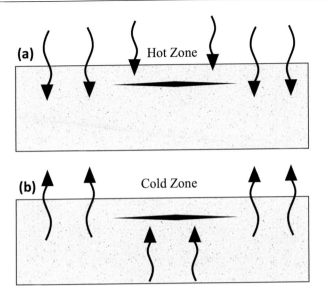

Figure 4.12 Effect of internal anomaly on surface temperature during heat flow. (a) Hot zone above flow due to inwards heat flow. (b) Cold zone above flaw due to outwards heat flow.

(Adapted from Malhotra & Carino 2003.)

investigations are necessary to improve the accuracy of outcomes and fully explore the potential of this technique when combined with appropriate data treatments.

4.3.3 Ground penetrating radar (GPR)

GPR is a geophysical method utilized to image sub-surfaces of structures by employing high-frequency electromagnetic waves. The antenna used in geophysical surveys serves the purpose of transmitting short pulses of electromagnetic energy within a specific frequency range (50–150 MHz) into the surveyed material. This antenna can be employed in two ways: it can either be dragged across the surface or attached to a survey vehicle. This technique capitalizes on the disparities in dielectric constants at material boundaries, where a portion of the energy is reflected while the remainder penetrates deeper into the structure. Figure 4.13 demonstrates a scheme of the GPR test.

GPR has gained widespread adoption as an NDT for assessing various aspects of concrete structures, including delamination, voids, reinforcing rebars and concrete thickness (ACI 201.1R 2008).

In the context of ISR-damaged infrastructures, GPR serves to complement VI data by assisting in the selection of coring locations and aiding in the classification of low- and high-damage regions within the structure. ISR

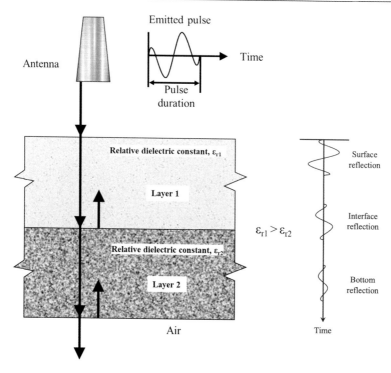

Figure 4.13 Reflections of electromagnetic radiation pulse at interfaces between materials with different relative dielectric constants.
(Adapted from Malhotra & Carino 2003.)

products exhibit distinct dielectric constants compared to concrete hydrated phases, which are discernible through variations in signal amplitude. Additionally, the time taken for the wave to bounce back to the receiver is utilized to determine the depth and location of the boundary, offering a somewhat accurate estimation of the ISR damage location, as observed in Figure 4.14.

GPR has proven to be an interesting auxiliary tool for assessing damaged concrete structures, enabling the estimation of regions impacted by ISR prior to the manifestation of visual signs. Nonetheless, further improvements in data treatment techniques and additional investigations are necessary to enhance the condition assessment of concrete infrastructure affected by ISR.

4.3.4 Stress waves

In isotropic and elastic media, stress waves propagate through two principal modes: dilatational waves, also known as compression waves or P-waves,

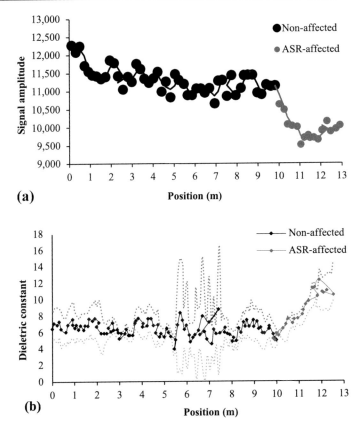

Figure 4.14 GPR treated outcomes (a) amplitude of dielectric constant and (b) dielectric constant.

(Adapted from Klysz et al. 2006.)

and distortional waves, known as shear waves or S-waves. These modes are distinguished by the direction of particle motion in relation to the propagation direction of the wavefront. Stress waves can be generated using various sources to evaluate damaged concrete, such as pulse-echo, impact-echo and impulse-response techniques (Malhotra & Carino 2003).

The most used method is the pulse echo, also known as ultrasonic pulse velocity (UPV). This test is performed by placing two transducers on the concrete surface in specific positions. These positions can be direct, semi-direct and indirect, as per Figure 4.15. One transducer will transmit a frequency, usually 54 kHz, and the other transducer will receive it; the more continuous the media, the higher the wave velocity. The wave mode depends on the transducer type applied.

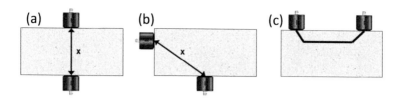

Figure 4.15 Schematic approaches to assess concrete through UPV: (a) direct, (b) semi-direct and (c) indirect.

P-wave analysis has been widely employed in the assessment of ISR-damaged concrete, particularly to appraise damage caused by ASR; its sensitivity is, however, limited, as demonstrated by Figure 4.16. The relationship between P-wave velocity and expansion varies depending on factors such as aggregate type (i.e., fine vs coarse), nature (i.e., lithotype) and amount of ASR-secondary products formed. Nonetheless, despite its limitations, the P-wave analysis is recommended in various protocols as a complementary tool for routine inspections (Godart et al. 2013).

Further research is, therefore, to explore the correlation between P-wave over ASR damage severity. This will contribute to refining the interpretation of P-wave data and enhancing the accuracy of ASR assessment using stress wave analysis. Additionally, investigating the effect of other parameters, such as specimen size and testing conditions, on the P-wave behaviour in ASR-damaged concrete would provide valuable insights for the development of more reliable assessment protocols.

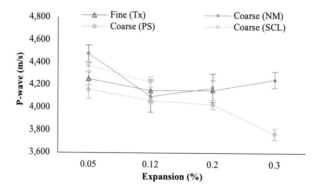

Figure 4.16 UPV (P-wave) in assessing ASR-affected concrete laboratory-made samples over expansion.

(Adapted from Rivard & Saint-Pierre 2009; Sanchez 2014; Sargolzahi et al. 2010.)

4.3.5 Resonant frequency (RF)

The natural frequency of vibration in an elastic system plays a crucial role in characterizing its dynamic behaviour since it is closely related to the dynamic modulus of elasticity and density of the system. By measuring the natural frequency of vibration, the dynamic modulus of elasticity of materials can be determined. In the assessment of concrete, the RF can be obtained through forced resonance tests and impact resonance tests (ASTM C 215 2019). RF analysis can be employed to evaluate ISR-affected concrete using both laboratory specimens or cores extracted from structures and real-scale members. The impact approach has been widely used to assess ISR, which can be obtained in three modes, as depicted in Figure 4.17.

The sensibility to detect ISR progression via RF, particularly for ASR-induced deterioration, has been observed to be influenced by several factors, such as aggregate's type (i.e., fine vs coarse), nature (i.e., lithotype), and amount of ASR-secondary products formed. Depending on the aggregate's lithotype, the reduction trend of the frequency can be observed in low ($\approx$ 0.05%) and moderate ($\approx$ 0.12) expansion levels. It is worth noting that no significant reduction in frequency is observed for ASR-affected concrete incorporating fine reactive aggregates, as illustrated in Figure 4.18. However, it is important to acknowledge that there are limitations to using RF as a sole indicator for assessing concrete distressed by ASR.

The assessment of real-scale concrete members affected by ASR using RF analysis has already been evaluated and yields valuable insights into the deterioration process. Specifically, the dynamic modulus of elasticity, as determined through the longitudinal frequency, exhibits a reduction in low-age specimens, indicating the sensitivity of RF to ASR progression at the real-scale level. However, according to (Siegert et al. 2005), it is important to note the presence of a rigidity restoration phenomenon attributed to a healing process which takes place over ASR development as per Figure 4.19. This restoration in rigidity is associated with the sealing of microcracks and pores filling by the ASR products, resulting in an overall improvement in the concrete stiffness. Furthermore, it should be acknowledged that structural effects, such as load redistribution and stress relaxation, along with environmental conditions (i.e., temperature, moisture, etc.), may also contribute significantly to the healing process observed in real-scale members.

Further research is still necessary to enhance the reliability and accuracy of RF analysis for ISR assessment. The investigation of the impact of various parameters, such as aggregate features, specimen size and geometry, and testing conditions on the RF responses is deemed crucial to enable the use of RF in a more diagnostic manner. Additionally, exploring the correlation between RF and other mechanical properties of ISR-affected concrete can contribute to a more comprehensive understanding of the deterioration process and aid in the development of effective evaluation methodologies.

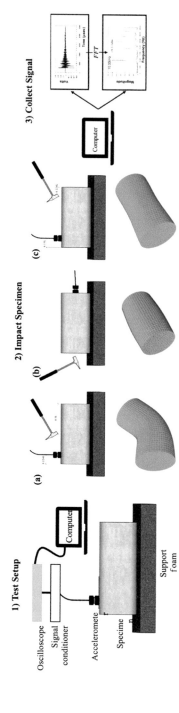

Figure 4.17 RF testing set-up for impact excitation: (a) transverse mode, (b) longitudinal mode and (c) torsional mode.

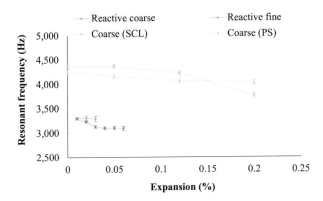

Figure 4.18 RF as a function of ASR damage degree.

(Adapted from Malone et al. 2021; Rivard & Saint-Pierre 2009; Sargolzahi et al. 2010.)

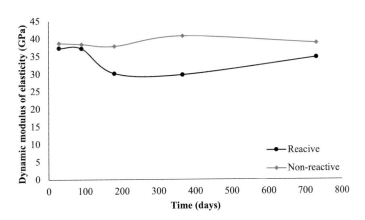

Figure 4.19 Dynamic modulus of elasticity over time of beams made of ASR-reactive and non-reactive aggregates.

(Adapted from Siegert et al. 2005.)

REFERENCES

ACI 201.1R. (2008). *Guide for conducting a visual inspection of concrete in service.* American Concrete Institute.

ACI 224R-19. (2019). *Control of cracking in concrete structures.* American Concrete Institute.

ASTM C 215. (2019). *Standard test method for fundamental transverse, longitudinal, and torsional resonant frequencies of concrete specimens.* ASTM International.

Auberg, R., & Setzer, M. (2014). *Frost resistance of concrete RILEM workshop on resistance of concrete to freezing and thawing with and without de-icing chemicals.* CRC Press.

Billah, U. H., Tavakkoli, A., & La, H. M. (2019). Concrete crack pixel classification using an encoder decoder based deep learning architecture. In *Advances in visual computing* (pp. 593–604). Springer International Publishing.

Carrasco, M., Araya-Letelier, G., Velázquez, R., & Visconti, P. (2021). Image-based automated width measurement of surface cracking. *Sensors, 21*(22). https://doi.org/10.3390/s21227534

Casanova, I., Agulló, L., & Aguado, A. (1996). Aggregate expansivity due to sulfide oxidation — I. Reaction system and rate model. *Cement and Concrete Research, 26*(7), 993–998. https://doi.org/10.1016/0008-8846(96)00085-3

Chinchón, J. S., Ayora, C., Aguado, A., & Guirado, F. (1995). Influence of weathering of iron sulfides contained in aggregates on concrete durability. *Cement and Concrete Research, 25*(6), 1264–1272. https://doi.org/10.1016/0008-8846(95)00119-W

Chopperla, K. S. T., & Ideker, J. H. (2022). Using electrical resistivity to determine the efficiency of supplementary cementitious materials to prevent alkali-silica reaction in concrete. *Cement and Concrete Composites, 125*, 104282. https://doi.org/10.1016/j.cemconcomp.2021.104282

David, L., & Gregory, M. (2017). Review of robotic infrastructure inspection systems. *The Journal of Infrastructure Systems, 23*(3), 1–12.

Dias, I. S., Flores-Colen, I., & Silva, A. (2021). Critical analysis about emerging technologies for Building's façade inspection. *Buildings, 11*(2), 1–19. https://doi.org/10.3390/buildings11020053

Duchesne, J., Rodrigues, A., & Fournier, B. (2021). Concrete damage due to oxidation of pyrrhotite-bearing aggregate: A review. *RILEM Technical Letters, 6*, 82–92. https://doi.org/10.21809/rilemtechlett.2021.138

Fasseu, P., & Michel, M. (1997). *Détermination de l'indice de fissuration d'un parement de béton; Méthode d'essai LCPC N0. 47.*

Fournier, B., Bérubé, M. A., Folliard, K., & Thomas, M. (2010). *Report on the Diagnosis, Prognosis, and Mitigation of Alkali-Silica Reaction (ASR) in Transportation Structures.*

Gattulli, V., & Chiaramonte, L. (2005). Condition assessment by visual inspection for a bridge management system. *Computer-Aided Civil and Infrastructure Engineering, 20*(2), 95–107.

Geiss, C. E., & Gourley, J. R. (2019). A thermomagnetic technique to quantify the risk of internal sulfur attack due to pyrrhotite. *Cement and Concrete Research, 115*, 1–7. https://doi.org/10.1016/j.cemconres.2018.09.010

Godart, B., & Divet, L. (2013). Lessons learned from structures damaged by delayed ettringite formation and the french prevention strategy. *Fifth International Conference on Forensic Engineering, Institution of Civil Engineers*, 389–400. https://hal.science/hal-00945667/document

Godart, B., Rooij, M., & Wood, J. G. M. (2013). *Guide to diagnosis and appraisal of AAR damage to concrete in structures: Part 1 diagnosis (AAR 6.1)* (RILEM). Springer.

Golden, J., Gomes, E., Anurag, R., & Su, Y. T. (2018). *Analysis of a historical failure, report no. 1: University of California, Berkeley.*

Grazia, M. T. (2023). *Short and long-term performance of eco-efficient concrete mixtures* [PhD]. University of Ottawa.

Gunn, R.M., Scrivener, K.L., & Leemann, A. (2017). The identification, extent and prognosis of alkali-aggregate reaction related to existing dams in Switzerland. In A. Sellier, É. Grimal, S. Multon, & É. Bourdarot (Eds.), *Swelling concrete in dams and hydraulic structures* (pp. 117–143).

Hansen, S. G., & Hoang, L. C. (2021). Anisotropic compressive behaviour of concrete from slabs damaged by alkali-silica reaction. *Construction and Building Materials, 267*. https://doi.org/10.1016/j.conbuildmat.2020.120377

Henrickson, J. V., Rogers. C., Lu, H. H., Valasek, J., & Shi, Y. (2016). Infrastructure assessment with small unmanned aircraft systems. *2016 International Conference on Unmanned Aircraft Systems (ICUAS)*, 933–942.

Jana, D. (2022). Cracking of residential concrete foundations in eastern Connecticut, USA from oxidation of pyrrhotite. *Case Studies in Construction Materials, 16*, e00909. https://doi.org/10.1016/j.cscm.2022.e00909

Jang, K., Kim, N., & An, Y. K. (2019). Deep learning–based autonomous concrete crack evaluation through hybrid image scanning. *Structural Health Monitoring, 18(5–6)*, 1722–1737. https://doi.org/10.1177/1475921718821719

Kabir, S. (2010a). Imaging-based detection of AAR induced map-crack damage in concrete structure. *NDT and E International, 43(6)*, 461–469. https://doi.org/10.1016/j.ndteint.2010.04.007

Kabir, S. (2010b). Imaging-based detection of AAR induced map-crack damage in concrete structure. *NDT & E International, 43(6)*, 461–469. https://doi.org/10.1016/j.ndteint.2010.04.007

Karthik, M. M., Mander, J. B., & Hurlebaus, S. (2016a). ASR/DEF related expansion in structural concrete: Model development and validation. *Construction and Building Materials, 128*, 238–247. https://doi.org/10.1016/j.conbuildmat.2016.10.084

Karthik, M. M., Mander, J. B., & Hurlebaus, S. (2016b). Deterioration data of a large-scale reinforced concrete specimen with severe ASR/DEF deterioration. *Construction and Building Materials, 124*, 20–30. https://doi.org/10.1016/j.conbuildmat.2016.07.072

Kim, H., Ahn, E., Shin, M., & Sim, S. (2019). Crack and Noncrack classification from concrete surface images using machine learning. *Structural Health Monitoring, 18(3)*, 725–738. https://doi.org/10.1177/1475921718768747

Kim, J. J., Kim, A. R., & Lee, S. W. (2020). Artificial neural network-based automated crack detection and analysis for the inspection of concrete structures. *Applied Sciences (Switzerland), 10(22)*, 1–13. https://doi.org/10.3390/app10228105

Klysz, G., Lataste, J.-F., Fnine, A., Dérobert, X., Piwakowski, B., & Buyle-Bodin, F. (2006). Auscultation non destructive du chevêtre du pont de la Marque (59). *Revue Européenne de Génie Civil, 10(1)*, 7–24. https://doi.org/10.1080/17747120.2006.9692813

Koch, C., Georgieva, K., Kasireddy, V., Akinci, B., & Fieguth, P. (2015). A review on computer vision based defect detection and condition assessment of concrete and asphalt civil infrastructure. *Advanced Engineering Informatics, 29(2)*, 196–210. https://doi.org/10.1016/j.aei.2015.01.008

Leemann, A., Menénendez, E., & Sanchez, L. (2021). Assessment of damage and expansion. In V. E. Saouma (Ed.), *Diagnosis & Prognosis of AAR Affected Structures State-of-the-Art Report of the RILEM Technical Committee 259-ISR*. Springer.

Liu, Y., Yeoh, J. K. W., & Chua, D. K. H. (2020). Deep learning–based enhancement of motion blurred UAV concrete crack images. *Journal of Computing in Civil Engineering, 34(5)*, 04020028. https://doi.org/10.1061/(asce)cp.1943-5487.0000907

Lugg, A., & Probert, D. (1996). 'Mundic'-type problems: A building material catastrophe. *Construction and Building Materials, 10(6)*, 467–474. https://doi.org/10.1016/0950-0618(95)00095-X

Mahadevan, S., Neal, K., Nath, P., Bao, Y., Cai, G., Orme, P., Adams, D., & Agarwal, V. (2017). Quantitative diagnosis and prognosis framework for concrete degradation due to alkali-silica reaction. *American Institute of Physics (AIP) Conference Proceedings, 1806*(080006). https://doi.org/10.1063/1.4974631

Malhotra, V. M., & Carino, N. J. (2003). *Handbook on nondestructive testing of concrete* (2nd ed.). CRC Press (Taylor & Francis Group).

Malone, C., Zhu, J., Hu, J., Snyder, A., & Giannini, E. (2021). Evaluation of alkali-silica reaction damage in concrete using linear and nonlinear resonance techniques. *Construction and Building Materials, 303,* 124538. https://doi.org/10.1016/j.conbuildmat.2021.124538

Martin, R. P., Céline, B., & Toutlemonde, F. (2015). *Alkali aggregate reaction and delayed ettringite formation: common features and differences. May 2012.* https://hal.archives-ouvertes.fr/hal-00852367

Miyamoto, A. (2013). Development of a remote collaborative visual inspection system for road condition assessment. *Key Engineering Materials, 569–570,* 135–142. https://doi.org/10.4028/www.scientific.net/KEM.569-570.135

Moliard, J.-M., Baltazart, V., Bérenger, B., Perrin, T., & Tessier, C. (2016). Digitized measurement of the cracking index on the facings of concrete structures. *8th RILEM International Conference on Mechanisms of Cracking and Debonding in Pavements, 13,* 731–737. https://doi.org/10.1007/978-94-024-0867-6

Morris, W., Moreno, E. I., & Sagüés, A. A. (1996). Practical evaluation of resistivity of concrete in test cylinders using a Wenner array probe. *Cement and Concrete Research, 26*(12), 1779–1787. https://doi.org/10.1016/S0008-8846(96)00175-5

Ogawa, N., Maeda, K., Ogawa, T., & Haseyama, M. (2022). Deterioration level estimation based on convolutional neural network using confidence-aware attention mechanism for infrastructure inspection. *Sensors, 22*(1). https://doi.org/10.3390/s22010382

Radopoulou, S. C., & Brilakis, I. (2017). Automated detection of multiple pavement defects. *Journal of Computing in Civil Engineering, 31*(2), 1–14. https://doi.org/10.1061/(asce)cp.1943-5487.0000623

Rivard, P., & Saint-Pierre, F. (2009). Assessing alkali-silica reaction damage to concrete with non-destructive methods: From the lab to the field. *Construction and Building Materials, 23*(2), 902–909. https://doi.org/10.1016/j.conbuildmat.2008.04.013

Rodrigues, A., Duchesne, J., Fournier, B., Durand, B., Rivard, P., & Shehata, M. (2012). Mineralogical and chemical assessment of concrete damaged by the oxidation of sulfide-bearing aggregates: Importance of thaumasite formation on reaction mechanisms. *Cement and Concrete Research, 42*(10), 1336–1347. https://doi.org/10.1016/j.cemconres.2012.06.008

Rodrigues, A., Duchesne, J., Fournier, B., Durand, B., Shehata, M. H., & Rivard, P. (2016). Evaluation protocol for concrete aggregates containing iron sulfide minerals. *ACI Materials Journal, 113*(3). https://doi.org/10.14359/51688828

Rogalski, A. (2011). Recent progress in infrared detector technologies. *Infrared Physics & Technology, 54*(3), 136–154. https://doi.org/10.1016/j.infrared.2010.12.003

Sanchez, L. F. M. (2014). *Contribution to the assessment of damage in aging concrete infrastructures affected by alkali-aggregate reaction,* PhD. 377. https://api.semanticscholar.org/CorpusID:139714980

Sargolzahi, M., Kodjo, S. A., Rivard, P., & Rhazi, J. (2010). Effectiveness of non-destructive testing for the evaluation of alkali-silica reaction in concrete. *Construction and Building Materials, 24*(8), 1398–1403. https://doi.org/10.1016/j.conbuildmat.2010.01.018

Savino, P., & Tondolo, F. (2021). Automated classification of civil structure defects based on convolutional neural network. *Frontiers of Structural and Civil Engineering*, 15(2), 305–317. https://doi.org/10.1007/s11709-021-0725-9

Siegert, D., Multon, S., & Toutlemonde, F. (2005). Resonant frequencies monitoring of alkali aggregate reaction (AAR) damaged concrete beams. *Experimental Techniques*, 29(6), 37–40. https://doi.org/10.1111/j.1747-1567.2005.tb00245.x

Strow, M., Bevington, P., Bentivegna, A., Bakhtiari, S., Aranson, I., Ozevin, D., & Heifetz, A. (2022). Monitoring accelerated alkali-silica reaction in concrete prisms with petrography and electrical conductivity measurements. *Materials and Structures*, 55(4), 119. https://doi.org/10.1617/s11527-022-01942-8

Thériault, F., Noël, M., & Sanchez, L. (2022). Simplified approach for quantitative inspections of concrete structures using digital image correlation. *Engineering Structures*, 252(June 2021), 1–12. https://doi.org/10.1016/j.engstruct.2021.113725

Thomas, M., Folliard, K., Drimalas, T., & Ramlochan, T. (2008). Diagnosing delayed ettringite formation in concrete structures. *Cement and Concrete Research*, 38(6), 841–847. https://doi.org/10.1016/j.cemconres.2008.01.003

Valença, J., Gonçalves, L. M. S., & Júlio, E. (2013). Damage assessment on concrete surfaces using multi-spectral image analysis. *Construction and Building Materials*, 40, 971–981. https://doi.org/10.1016/j.conbuildmat.2012.11.061

Vlahović, M. M., Savić, M. M., Martinović, S. P., Boljanac, T. D., & Volkov-Husović, T. D. (2012). Use of image analysis for durability testing of sulfur concrete and Portland cement concrete. *Materials and Design*, 34, 346–354. https://doi.org/10.1016/j.matdes.2011.08.026

Wang, Y., Ramanathan, S., Chopperla, K. S. T., Ideker, J. H., & Suraneni, P. (2022). Estimation of non-traditional supplementary cementitious materials potential to prevent alkali-silica reaction using pozzolanic reactivity and bulk resistivity. *Cement and Concrete Composites*, 133, 104723. https://doi.org/10.1016/j.cemconcomp.2022.104723

Wood, J. G. M. (2008). Implications of the collapse of the de La Concorde overpass. *4th International Conference on 'Forensic Engineering'*, 16–18.

Zahedi, A., Sanchez, L.F.M., & Noël, M. (2022). Appraisal of visual inspection techniques to understand and describe ASR-induced development under distinct confinement conditions. *Construction and Building Materials*, 323, 126549. https://doi.org/10.1016/j.conbuildmat.2022.126549

Zhang, A., Wang, K. C. P., Fei, Y., Liu, Y., Tao, S., Chen, C., Li, J. Q., & Li, B. (2018). Deep learning–based fully automated pavement crack detection on 3D asphalt surfaces with an improved CrackNet. *Journal of Computing in Civil Engineering*, 32(5), 1–14. https://doi.org/10.1061/(asce)cp.1943-5487.0000775

Zhang, A., Wang, K. C. P., Li, B., Yang, E., Dai, X., Peng, Y., Fei, Y., Liu, Y., Li, J. Q., & Chen, C. (2017). Automated pixel-level pavement crack detection on 3D asphalt surfaces using a deep-learning network. *Computer-Aided Civil and Infrastructure Engineering*, 32(10), 805–819. https://doi.org/10.1111/mice.12297

Zhong, R., & Wille, K. (2018). Deterioration of residential concrete foundations: The role of pyrrhotite-bearing aggregate. *Cement and Concrete Composites*, 94, 53–61. https://doi.org/10.1016/j.cemconcomp.2018.08.012

Zhu, Z., German, S., & Brilakis, I. (2011). Visual retrieval of concrete crack properties for automated post-earthquake structural safety evaluation. *Automation in Construction*, 20(7), 874–883. https://doi.org/10.1016/j.autcon.2011.03.004

Chapter 5

Microscopic analyses

5.1 INTRODUCTION

As previously discussed in Chapter 4, damage at the surface of concrete detected through visual inspection (VI) methods (i.e., qualitative descriptions and cracking index – CI) and non-destructive testing (NDT) can give a first glance at its condition. Mechanical testing subsequently corresponds to the material's macro performance and ability to continue withstanding loads (see Chapter 6). Although the exact cause of damage may not necessarily be captured by VI, NDT and certain mechanical tests, it is widely accepted that a combination of microscopy tools (i.e., qualitative and quantitative) can offer insight into the cause(s) leading to concrete deterioration (e.g., internal swelling reaction (ISR) type) along with the damage degree of the affected material. This chapter intends to display the common qualitative and quantitative microscopic procedures used to assess the cause and extent of ISR damage in concrete in both micro and mesoscales.

5.2 PETROGRAPHIC ANALYSIS: ASSESSING THE DAMAGE CAUSE(S)

5.2.1 Introduction

Petrographic analysis refers to the study of concrete (or rock) thin or polished concrete sections, allowing a petrographer to reveal certain characteristics of the concrete (or rock) microstructure, such as aggregates mineralogy, various phases of cement hydration and secondary reaction products. For many years, the science of petrography has played an important role in concrete research, especially in the condition assessment of damaged concrete (Jana, 2005). Over time, petrography became widely used in the concrete construction industry with the aim of

a. characterizing concrete's composition, mineralogical and textural properties, including aggregate lithotypes, size and shape, cement paste features, presence of supplementary cementing materials and chemical admixtures

DOI: 10.1201/9781003188155-5

(Erlin & Stark, 1990) as per the *Standard Guide for Petrographic Examination of Aggregates for Concrete* (ASTM C295, 2019);

b. providing a quality control tool for concrete construction (i.e., incorrect proportioning, mixing, placement, finishing or curing operations) as per the *Standard Practice for Petrographic Examination of Hardened Concrete* (ASTM C856-20, 2020);

c. diagnosing the cause(s) of concrete damage due to deleterious chemical (e.g., acid, alkali, sulphate, chloride, seawater attack) and physical (e.g., frost, fire) mechanisms (Erlin & Stark, 1990; A. B. Poole & Sims, 2016; Walker et al., 2006); and

d. appraising the efficiency of rehabilitation strategies (e.g., repair materials).

Petrographic evaluations are generally conducted using optical microscopes (OM), scanning electron microscopes (SEM) or both (ASTM C295, 2012; ASTM C856-20, 2020; ASTM C1723-16, 2016; A. B. Poole & Sims, 2016; Walker et al., 2006). Chemical compositions of distinct concrete microstructure phases can also be determined using energy-dispersive X-ray spectroscopy (EDS/EDX). Petrography should be conducted by highly qualified, skilled and experienced petrographers as per American standards to be reliable (ASTM C295, 2019; ASTM C856-20, 2020). Moreover, it is suggested that a petrographer be familiar with the local geology of the concrete structure under evaluation (Fernandes et al., 2016).

5.2.2 Sample preparation for petrography

Concrete specimens, either cores extracted from concrete structures or structural members or even manufactured in the laboratory, are sized to adequately represent the region of interest for analysis. Initially, cores are cleaned after their extraction and then wrapped in self-adhesive film and/or sealed bags to prevent dehydration and further deterioration (Broekmans, 2012; CUR-Recommendation-102, 2008). The specimens are then cut to size (i.e., thin or polished sections), cleaned and dried. Afterwards, the sections are impregnated with a clear or fluorescent dye epoxy (ASTM C856-20, 2020; Broekmans, 2012; CUR-Recommendation-102, 2008; Poole & Sims, 2016) to create a contrast using ultraviolet (UV) lights where the resin is deposited, such as in voids, porous media and cracks. Finally, the samples are further ground/polished to achieve minimal topographical variations and light scattering, thus enhancing the distinction between concrete components under SEM and X-ray microanalyses (Goldstein et al., 2003).

Figure 5.1 illustrates conventional (i.e., 50 mm by 28 mm) and large-size (i.e., 50 mm by 100 mm) thin sections (A and B, respectively) along with mounted and impregnated polished concrete sections (C and D) prepared for petrographic analysis. Several guides are readily available for further reference on aggregate and concrete petrography (Fernandes et al., 2016; Ingham, 2013; Poole & Sims, 2016).

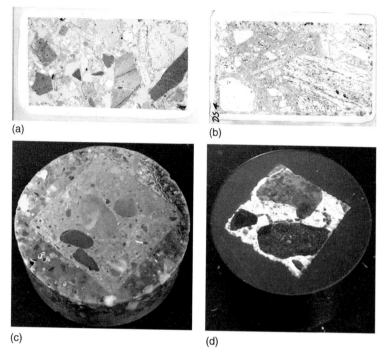

(a)   (b)

(c)   (d)

Figure 5.1 Images taken with a flatbed scanner of (a) conventional size thin section of 50 mm by 28 mm, (b) large thin section of 50 mm by 100 mm, (c) 4 cm diameter concrete section and (d) 2.5 cm diameter mounted concrete section.

(Photographs courtesy of Andreas Leemann, Leandro Sanchez, Cassandra Trottier.)

5.2.3 Evaluation of ISR in concrete through petrography

The first step in the evaluation of ISR-induced deterioration in concrete is the selection of regions of interest (ROI) in the sample under analysis (ASTM C856-20, 2020); normally, this selection is made via a thorough appraisal using the stereomicroscope. Figure 5.2 illustrates a stereomicroscope and a prepared specimen with highlighted ROI.

Once the ROI is defined, an optical microscope is used to detect the presence of reactive minerals within the aggregates, reaction products, mixture proportions (estimated range of water-to-cement ratio, air-void characteristics), etc. This appraisal is normally conducted through cross-polarized transmitted light or reflected light at magnifications up to 600x (Jana, 2005). Complimentary evaluations under the SEM with EDS/EDX might be performed; the latter allows for microscopic examinations at higher magnifications, along with providing chemical composition analysis. Figure 5.4a and

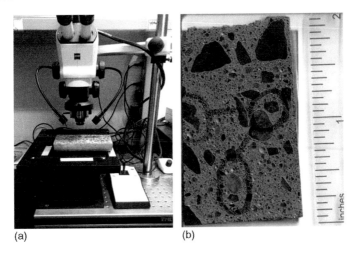

(a)   (b)

Figure 5.2 Stereomicroscope equipped with an automated stage for automatic image acquisition and linear traverse stepwise operations used for preliminary analysis and selection of ROI on polished concrete sections.

(Photograph courtesy of Francisco Locati and Cassandra Trottier.)

b illustrate secondary reaction products of ettringite and alkali-silica reaction (ASR) with their corresponding EDS/EDX, respectively. For example, ettringite may have the appearance of needle-like crystals filling or lining empty spaces in the concrete microstructure, presenting silica, alumina, and calcium as the main oxides (Melo et al., 2023; Thomas et al., 2008). Otherwise, ASR secondary products (i.e., the so-called ASR "gel") may display various morphologies (i.e., amorphous or crystalline) and varied chemical composition according to its age and location in the concrete microstructure (higher amounts of calcium are verified in regions closer to the cement paste); nevertheless, silica, sodium, potassium, and calcium are the main oxides present in ASR secondary products (Leemann et al., 2016). Ultimately, to confirm whether the products encountered via EDS/EDX are the ones suspected, their atomic mass percentages may be evaluated through ratio plots, as illustrated in Figure 5.3a and b, in the case of secondary ettringite and ASR, respectively. If the ratios measured via EDS/EDX are close to (or move towards) the ettringite (Ett.) dot in Figure 5.4a or show total alkalis to silica ratio (i.e., Na+K+Cs/Si) values between 0.20 and 0.35 as per Leemann et al. and Ahmed et al., 2022 in Figure 5.4b, then these products are confirmed to be secondary ettringite or ASR products, respectively. The previous discussion highlights the great importance and reliability of petrographic analysis to assess the cause(s) of deterioration of concrete infrastructure; nevertheless, the following limitations still remain during the use of this technique: (a) high expertise is required to properly prepare

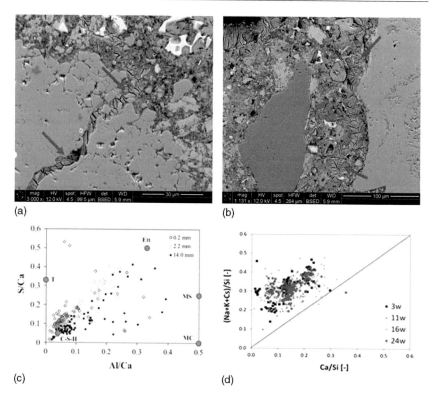

Figure 5.3 SEM images of (a) ettringite-filled crack in the aggregate (shown with solid arrows) extending into the cement paste (shown with dashed arrow), (b) ettringite lining a void at the top left corner (shown with dashed arrow) and surrounding an aggregate particle (shown with solid arrows), (c) EDS/EDX analysis showing the presence of ettringite (Loser & Leemann, 2016) and (d) EDS/EDX analysis showing the presence of ASR secondary products.

(Figure courtesy of Leandro Sanchez).

specimens, conduct the analysis and evaluate outcomes to reduce subjectivity; (b) the specimen size assessed may not be representative to explain the whole deterioration process observed in the concrete structure or structural member under analysis; (c) although reliable to assess the cause(s) leading to deterioration, conventional petrography provides limited quantitative information on the extent of deterioration experienced by the affected concrete. Such information is crucial to help engineers and infrastructure owners make better decisions in terms of management and rehabilitation protocols (Hollis et al., 2006; Poole & Sims, 2016). In this context, a number of microscopic techniques have been proposed over the last years to address the aforementioned drawbacks, and they will be discussed in the following section.

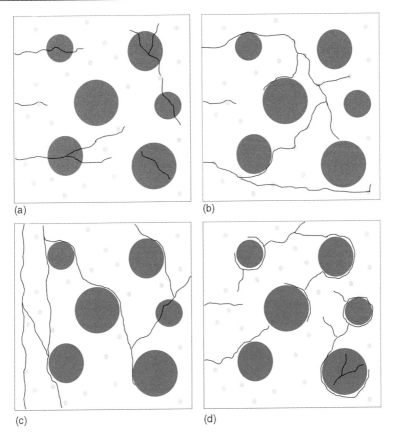

Figure 5.4 Crack pattern in concrete affected by (a) ASR reactive coarse aggregate, (b) ASR reactive sand, (c) FT cycles and (d) DEF. The concrete's surface is shown on the left-hand side.

(Adapted from British Cement Association (BCA), 1992.)

5.3 PETROGRAPHIC ANALYSIS: ASSESSING THE DAMAGE DEGREE

5.3.1 Introduction

Understanding the petrographic distress features and degree of concrete is a matter of scale; depending on the evaluation size, different microstructure features (i.e., cracks, defects, flaws, etc.) are observed, yet the analysis scale should be selected as per the examiner's purpose. In Chapter 3, damage was globally defined as (1) engineering properties reductions (i.e., compressive, tensile and direct shear strengths) of concrete, (2) stiffness reduction of concrete and (3) physical integrity and/or durability loss, which is directly related to the existing crack network. Therefore, whether or not petrographic analysis is to be used to evaluate damage, it should be able to correlate the observed petrographic distress features with the previously described damage parameters. In this context, the literature demonstrates that petrographic distress features observed at the mesoscale (i.e., polished concrete sections at about 15x–20x magnification) may properly correlate with macro performance when compared to standardized mechanical and durability testing protocols (Sanchez et al., 2017, 2018). Furthermore, at the mesoscale, different damage patterns may be observed as per the deterioration mechanism(s) leading the overall damage process, which may not only facilitate the detection of the may cause(s) of damage but actually help understand the potential impact on the engineering properties and durability of the material. For instance, Figure 5.4 illustrates common petrographic damage features and patterns observed for distinct ISR mechanisms. For ASR-induced deterioration, cracks originate and are mainly found within the aggregate particles, either coarse (Figure 5.4a) or fine (Figure 5.4b). Otherwise, both freeze and thaw (FT) and DEF induce cracks in the cement paste, where FT-induced cracks are present in the bulk cement paste and pores, being fairly parallel to the concrete's surface in locations exposed to the environment, while DEF-induced cracks are more located in the ITZ and connect to one another in the cement paste at high deterioration levels. A number of petrographic protocols at the mesoscale have been developed over the past decades to semi or fully quantitatively appraise deterioration in concrete. Amongst them, the *Damage Rating Index (DRI)*, a semi-quantitative petrographic procedure and the *image analysis*, a fully quantitative approach, seem to be the most effective. The next section will present these methods along with some emerging technologies to quantitatively evaluate ISR-induced deterioration in concrete.

5.3.2 Damage rating index (DRI)

The DRI method consists of analysing 1 cm^2 grids (field of view at 15x–16x magnification) defined at the surface of polishing concrete sections using a stereomicroscope (reflected light). In each square, various types of petrographic damage features are counted and multiplied by weighting factors whose main purpose is to balance their relative importance towards the overall deterioration of the concrete. The previous calculation (i.e., number of cracks multiplied by the respective weighting factors) corresponds to the DRI number; the higher the DRI number, the higher the deterioration of the concrete. Ideally, a surface of at least 200 cm^2 should be used for analysis, and it may be greater in the case of mass concrete incorporating coarse aggregates with larger size fractions. Nonetheless, for comparative purposes, the final DRI number is normalized to a 100 cm^2 area (Figure 5.5).

The DRI was initially idealized by (Grattan-Bellew & Danay, 1992) for assessing the damage caused by ASR; therefore, the first weighting factors were proposed to indicate the level of development, or progress, of ASR in

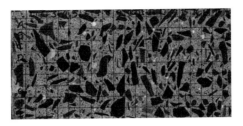

Figure 5.5 A prepared specimen for DRI analysis.

(Photograph courtesy of Cassandra Trottier.)

concrete (Grattan-Bellew & Danay, 1992). Over time, the weighting factors were modified so that the DRI number could better correlate with induced expansion, physical integrity and engineering properties reductions of ISR-affected concrete (Sanchez et al., 2015, 2017, 2018, 2020; Villeneuve, 2011). The types of cracks appraised in the DRI method are described in Table 5.1, along with their respective weighting factors. It is worth noting that higher weighting factors are attributed to features deemed more important. For instance, a closed crack in the aggregate may not represent ISR damage, as it is most likely produced by rock weathering and aggregate processing; hence, a factor below unity (i.e., 0.25) is attributed to this feature. Otherwise, an opened crack in the aggregate (with or without secondary products) can indicate that a reactive aggregate is present or even damage in the aggregate has occurred due to pressure from its surroundings; a factor of 2 is attributed to such cracks. One should note that secondary products can be washed away during polishing, besides being difficult to observe at 15x–16x magnification. Thus, the same factor has been attributed to open cracks in the aggregates with and without products. Moreover, a disaggregated/corroded particle has the same weight as cracks within aggregate particles since the damage remains within the particle and is of rare occurrence yet may indicate the presence of ASR.

Cracks in the cement paste represent the highest level of damage with a weighting factor of 3. Such cracks indicate either the presence of mechanisms generating and propagating damage in the cement paste, such as DEF and FT, or the presence of a well-developed ASR, where its cracks have elongated from the aggregate particles into the cement paste. Like the reaction products within the aggregate particles and for the same reasons, the cement paste cracks with and without reaction products are weighted the same. Finally, the debonded aggregate feature, which also bears a weighting factor of 3, is quite common for mechanisms generating cracks in the cement paste, especially DEF, while remaining a rare occurrence in ASR-affected concrete.

The DRI has been shown over the last years to be a reliable tool to appraise damage caused by ISR in concrete (Sanchez et al., 2018, 2020).

Table 5.1 The DRI weighting factors associated with ISR damage

Distress feature	Description	Acronym	Weighting factor
Closed crack in aggregate	A crack in which no opening is observed	CCA	0.25
Opened crack in aggregate	A crack with an opening between its edges	OCA	2
Opened crack in aggregate with reaction product	A crack with an opening between its edges and filled with a substance	OCA-RP	2
Disaggregated/ corroded aggregate particle[b]	A reacted particle in which its centre has been eroded and removed	DAP	2
Crack in the cement paste	An open crack propagating through the cement paste	CCP	3
Crack in the cement paste with reaction product	An open crack filled with a substance propagating through the cement paste	CCP-RP	3
Coarse aggregate debonding[a]	A crack highlighting the aggregate-paste interface	CAD	3

[a] A debonded aggregate is considered when a crack in the ITZ surrounds 50% of an ASR reactive particle. When DEF is suspected, a debonded aggregate is considered when the crack in the ITZ surrounds 75% of the aggregate.
[b] A disaggregated/corroded particle must not contain a countable crack network.

However, its non-negligible subjective character, which may cause variability in the test outcomes, is still criticized by engineers and practitioners. Amongst the aspects that may potentially bring important subjectivity and variability to the method, the distinct crack features recognition and count are the most important. Otherwise, like conventional petrography, the ability to recognize different crack features comes with practice and training. Figure 5.6 shows a few of the distress features observed through a stereomicroscope at 16x magnification. It is important to note that the count of cracks may be quite complicated, especially in systems containing important crack networks, such as illustrated in Figure 5.7. In these systems, a node and segment method may be used to count the cracks. However, other techniques may also be used, such as defining a starting point (i.e., reference) and following the cracks from this point onwards until their ends; both methods result in the same counts. In Figure 5.7, one verifies the node and segment calculation where the colour coded lines reveal distinct cracks (i.e., segments: 1-2-3, 9-6, 4-5, 7-8, 10-11-12, 13, 15 and 14-16; nodes: 1 to 8).

Once all cracks observed on the sample are counted, they are summed and normalized to 100 cm^2, as per Equation 5.1. It is suggested to electronically tabulate the counts using software to calculate the final DRI number and produce the corresponding bar charts, illustrated in Figure 5.8.

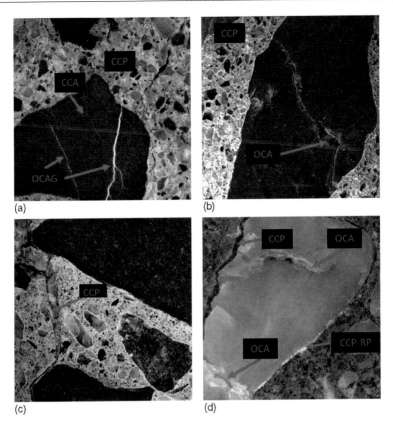

Figure 5.6 Micrographs of I cm² presenting distress features counted through the DRI of ISR in concrete: (a) concrete affected by ASR with open cracks in the aggregate with gel (OCAG), closed crack in the aggregate (CCA) and crack in the cement paste (CCP), (b) concrete affected by alkali-carbonate reaction with an open crack in the aggregate (OCA) and a crack in the cement paste, (c) concrete affected by FT with cement paste crack (CCP) parallel to the exposed surface and (d) DEF affected concrete with open cracks in the aggregate (OCA) and cracks in the cement paste with and without reaction product in the ITZ.

(Photographs courtesy of Andisheh Zahedi and Cassandra Trottier.)

$$\text{DRI} = \left[\frac{\sum \left(0.25(CCA) + 2(OCA + OCA_{RP} + DAP) + 3(CCP + CCP_{RP} + CAD) \right)}{\text{Number of analysed 1 cm by 1 cm squares}} \right] \times 100\ \text{cm}^2$$

$$(5.1)$$

The bar chart in Figure 5.8 illustrates not only the DRI number (i.e., x-axis) but also the distinct crack features (i.e., distinct colours) observed,

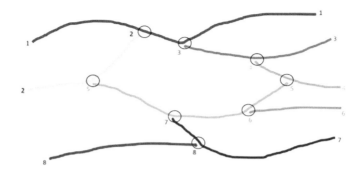

Figure 5.7 Point count method for crack networks.

(Adapted from Sims et al., 1992.)

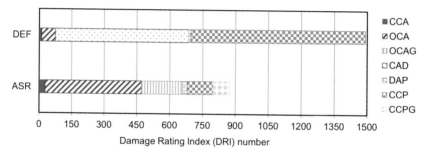

Figure 5.8 DRI bar chart comparing ASR and DEF.

which gives a better understanding of the mechanism(s) leading to the deterioration of the affected concrete. For example, DEF generates an important number of cracks in the cement paste (CCP – checker chart), several of them causing debonding of the aggregates from the cement paste (CAD – dotted chart). Otherwise, ASR generates cracks within the aggregate particles (OCA – diagonal striped chart), some of them presenting secondary products (OCA$_{RP}$ – vertical stripe chart); upon development, some of these cracks elongate to the cement paste, generating cracks with (CCP$_{RP}$ – diamond chart) and without (CCP – checker chart) reaction products.

Other than the DRI bar charts, the DRI can be performed in its extended version (Sanchez et al., 2015). In this extended method, a thorough analysis of the distinct petrographic damage features is performed in both counts/100 cm^2 and percentages without the use of the weighting factors. Figure 5.9 illustrates plots in counts (Figure 5.9a) and percentages (Figure 5.9b). These visual representations help to distinguish the leading crack types, which may be composed of (i) CCA, (ii) OCA+OCA$_{RP}$, (iii) CCP+CCP$_{RP}$ and (iv) CAD. In addition, the crack density as the summation of OCA+OCA$_{RP}$ and CCP+CCP$_{RP}$ counts per 1 cm^2 can be plotted as a function of expansion (Figure 5.9c).

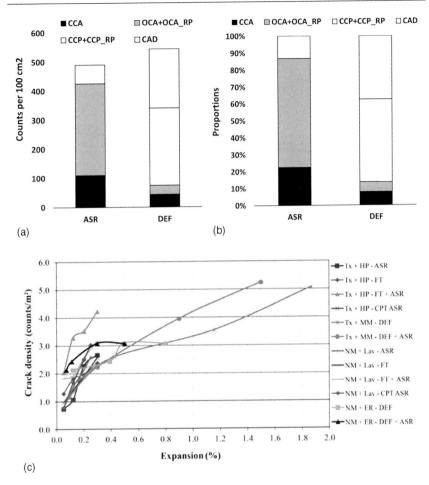

Figure 5.9 Extended version of the DRI with (a) crack counts per 100 cm², (b) proportions and (c) crack density.

(Sanchez et al., 2020).

The annotation, description and record of other petrographic features that are observed while gathering the crack counts for the DRI calculation is highly recommended to supplement the investigation into the cause of the damage. These features can include but are not limited to the deposit of product in voids, reaction rims where a dark rim is observed outlining the aggregate particles, the crack pattern, the identification of extensions such as cracks extending from aggregate to cement paste and so on. Some of the features may not necessarily represent damage to concrete but can indicate the presence of ISR.

5.3.2.1 Sample preparation for DRI

Sample preparation for the DRI procedure is fairly simple and follows steps that are common for microscopy. Initially, a concrete core or specimen is cut in half longitudinally using a masonry saw equipped with a diamond blade, either unnotched or notched (Figure 5.10a). The masonry saw can include a water kit or other coolants, such as paraffin oil or kerosene; care must be taken when using coolants other than water and all safety protocols followed. It is suggested to cut the concrete core or specimen in one single motion to reduce deep saw marks on the finished surface (Figure 5.10b). The cut is then followed by a series of subsequent grinding/polishing stages to achieve a flat and highly reflective surface. A suggested sequence of lapping disks from coarsest to finest is as follows: 30 grit (to remove saw marks if necessary), 60 grit, 140 grit, 280 grit (80–100 μm), 600 grit (20–40 μm), 1,200 grit (10–20 μm) and 3,000 grit (4–8 μm). However, this series is based on the supplier and type of abrasive material used. Figures 5.11a and b

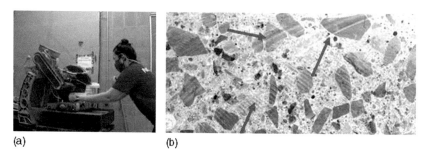

(a)   (b)

Figure 5.10 DRI sample cutting: (a) longitudinal cutting of concrete specimen using a wet masonry saw equipped with a 14-inch diamond blade and (b) saw marks visible on the surface after cutting.

(Photographs courtesy of Cassandra Trottier.)

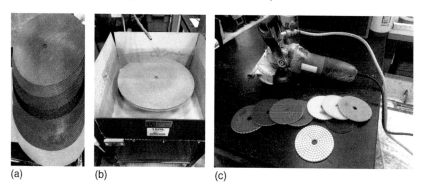

(a)   (b)   (c)

Figure 5.11 (a) 18-inch magnetic diamond-impregnated laps, (b) mechanically rotating lapping machine and (c) handheld polisher.

(Photographs courtesy of Cassandra Trottier.)

show 18-inch lapping disks and a mechanically rotating wheel, respectively, while Figure 5.11c shows a handheld mechanical polisher. A glass plate can also be used onto which abrasive powders are scattered (Figure 5.12a) and mixed with a lubricant such as water to produce a grinding/polishing paste (Figure 5.12b). Manual random circular motions are then used while applying some pressure and ensuring that the entire surface is ground/polished (Figure 5.12c). The time required for each step is dependent on the equipment/method used and the types of materials comprised in the concrete. In any case, the surface is cleaned (using a soft bristle brush or compressed air) after each step to remove the slurry (Figure 5.12d), any loose abrasives or aggregate/cement paste fragments that may have become dislodged during sample preparation. A quick examination of the surface quality after each step can be performed by reflecting sunlight or indoor ceiling lights onto the surface to observe if any portion of the surface lacks reflectiveness, as shown in Figure 5.15a, where the centre portion of the surface is duller and whiter than the sides. Moreover, deep scratches or marks from previous steps visible on the aggregates with the naked eye should be removed prior to the polishing stages. Marks from previous steps should be progressively less visible on the aggregates as the surface becomes ready for the next step. If the aggregates have the appearance of being scuffed when using impregnated disks, it is an indication that more lubricant, such as water, is required during the polishing step.

When the surface of the section is considered suitable for analysis, having a highly reflective surface (Figure 5.13b), a grid of 1×1 cm squares is drawn on the surface using a thin tip permanent marker; an alternative to drawing and permanently marking the surface of the polished concrete section under analysis is the use of a 3D printed grid that is simply placed on the specimen's surface, as shown in Figure 5.13c. Stiff gardening mesh with a 1 cm by 1 cm grid can also be used. Generally, the first row is used to number the

(a)　　　　　(b)　　　　　(c)　　　　　(d)

Figure 5.12 Abrasive powder (a) scattered onto a wet glass plate, (b) mixed by hand to form a slurry, (c) placing of the concrete specimen with its working surface against the glass plate and (d) the excess of slurry onto the working surface prior to cleaning.

(Photographs courtesy of Cassandra Trottier and Francisco Locati.)

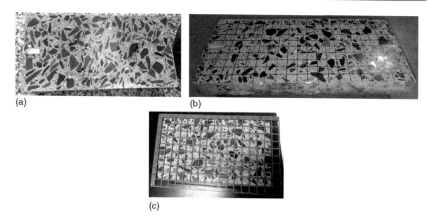

Figure 5.13 A concrete surface (a) not well finished where a non-reflective area is observed at the centre of the surface shown as a whiter colour compared to both sides, (b) a well-finished, highly reflective surface with a grid drawn where light reflection is visible in the bottom right corner and numbering convention shown on the first row and column and (c) with a 3D-printed grid placed over the finished surface. **(Photographs courtesy of Cassandra Trottier.)**

columns (for example, 1 to 20, as Figure 5.13b illustrates), while the first column is used to label the rows (for example, A to I, as Figure 5.13b illustrates).

5.3.3 Image analysis

Image analysis (or digital image processing – DIP) can also be applied to quantify damage in concrete. Once captured, images are analysed either manually or using software where quantitative information pertaining to the deterioration characteristics is extracted, such as the crack length, width and pattern (i.e., isotropic vs anisotropic), amongst many others often observed in 2D images (a discussion on 3D images is presented in Section 5.4). Cracks characteristics such as location, orientation and pattern with respect to (a) the loading direction of the structural member and (b) the exposed surface and the presence of reaction products are to be considered when identifying the cause(s) and extent of damage.

Image analysis is normally used, as the DRI, after conventional petrography, where the main cause(s) leading to deterioration is found. Impregnation is recommended using a fluorescent dye to highlight the cracks encountered and thus facilitate precise quantification of the petrographic damage features (providing greater contrast between cracks, voids and flaws and the background). Figure 5.14 illustrates concrete polished samples impregnated with a fluorescent dye, where Figure 5.14a shows an ASR specimen presenting a

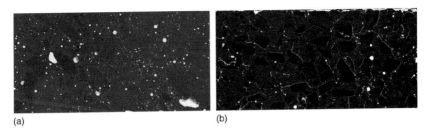

(a)   (b)

Figure 5.14 Fluorescent epoxy impregnated 5 cm by 10 cm concrete section illuminated by UV lights: (a) low ASR-induced expansion specimen (i.e.,
0.05%) and (b) high ASR-induced expansion specimen (i.e., 0.20%).

(Photograph courtesy of Cassandra Trottier and Leandro Sanchez.)

low deterioration, and Figure 5.14b illustrates an ASR sample displaying a
high deterioration.

The outcomes of image analysis, such as crack length, width, pattern,
orientation and density, can be associated with (a) the level of damage and
(b) the impact of the observed cracks on the mechanical properties, stiffness
and durability of the ISR-affected concrete. Crack lengths and orientation
were found to be the best computable indicators to estimate induced expansion caused by ASR when evaluating laboratory and field specimens (Rivard
et al., 2000). Meanwhile, the "Norwegian crack index" applied to ASR-
affected concrete where a plane polished surface is impregnated with fluorescent material and illuminated with UV light without using a microscope,
in which the proportions of the cracks in the aggregate particles and cracks
extending into the cement paste from the aggregates are counted and
summed over 100 cm^2, was used to estimate ASR-induced expansion
(Jensen & Sujjavanich, 2016; Lindgård et al., 2004, 2012).

5.3.3.1 Sample preparation for image analysis

The sample preparation for image analysis consists of applying a fluorescent
dye to enhance the appearance of cracks under UV light. Specimens are cut
axially and ground flat. The fluorescent epoxy resin is then applied to the
flat dry surface (Figure 5.15a), placed under a vacuum to ensure that cracks
are filled (Figure 5.15b) and left to cure as per the manufacturer's details.
Figure 5.15c shows concrete sections with hardened/cured fluorescent
epoxy prior to it being removed to expose the cracked surface. Once the
coat has hardened, it is carefully removed by following a similar grinding/
polishing sequence as for the DRI while ensuring that the resin-filled cracks
remain at the surface without removing the impregnated cracked material.
If cracks are previously filled with reaction products, it may not be feasible
to fill such cracks with resin, and therefore, this should be considered during the analysis. An important note when using fluorescence is to ensure

(a)   (b)   (c)

Figure 5.15 Fluorescent epoxy resin (a) being applied to a concrete section, (b) vacuum chamber and (c) hardened/cured surfaces ready for removal. **(Photographs courtesy of Cassandra Trottier and Francisco Locati.)**

that the proportions of fluorescent powder and resin are compatible with the UV light source. Trials may be required to determine the proportions unless a pre-mixed fluorescent resin is used. A flatbed scanner can further be used for image analysis as an effective approach to acquire an intact image (i.e., without stitching several images) and has been previously used to evaluate the amount of entrained air in concrete (Fonseca & Scherer, 2015; Song et al., 2017). An example of an image taken with a flatbed scanner is shown in Figure 5.5.

5.4 EMERGING TECHNOLOGIES TO QUANTIFY DAMAGE IN CONCRETE

Evidently, the digitization of images used for image analysis leads to automating the manual classification of petrographic damage features and the mathematical interpretation of extracted data. Although many of the procedures of image analysis are automated to filter the images and remove noise and unwanted features, such as voids, stitching and thresholding, amongst many other tedious tasks, very few automated protocols have been developed to quantitatively assess ISR-induced deterioration in concrete (Rivard, 1998). Dehghan et al. (2016) used the DRI on digital images taken with two methods: (a) a flatbed scanner and (b) a stereomicroscope equipped with a digital camera, which is a form of image analysis in a point count or segmentation sense. However, with the rise in new technologies, more sophisticated image acquisition equipment and availability of previously written scripts, it becomes less of a challenge to automate image analysis procedures. Some works have used artificial intelligence to identify cracks and classify them based on the DRI's classification system (Bezerra, 2020), which shows a clear contrast more easily detectable when compared to fluorescence, as shown in Figure 5.16. Moreover, since the arrival of artificial intelligence techniques,

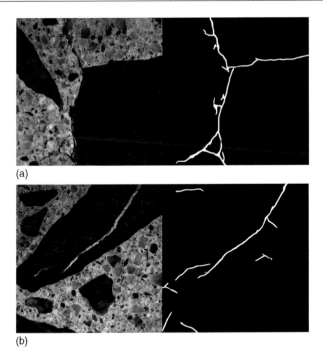

Figure 5.16 Micrographs of 1 cm² polished section from a Robert Bourassa-Charest overpass extracted core on the left-hand side and traced cracks using an automated procedure (Bezerra, 2020) on the right-hand side: (a) shows CCP and ITZ, while (b) shows cracks mainly in the aggregate with the reaction product.

(Photos courtesy of Agnes Bezerra, Haoye Lu and Cassandra Trottier.)

object recognition in images has been continuously improving, which could potentially replace the time-consuming sample preparation used to enhance the contrast between the cracks and other components of the concrete (i.e., cement paste, voids and aggregate) while reducing the potential for induced shrinkage cracks often observed after drying the concrete specimens before application of fluorescent dyes.

Three-dimensional image analysis using micro-computed tomography (micro-CT) has begun to gain more interest in the diagnosis of ISR-affected concrete, particularly to measure cracks generation and propagation, along with reaction product formation. As such, a number of researchers have used micro-CT to evaluate ASR in mortar specimens (Hernández-Cruz et al., 2016; Marinoni et al., 2009; Voltolini et al., 2011), whereas others have evaluated the progression of ASR in concrete over time (Shakoorioskooie et al., 2021). Moreover, Suzuki et al. (2017) CT scan appraised cores extracted from concrete columns affected by FT to estimate the induced deterioration in correlation with acoustic emission. Likewise (Joshi et al., 2022), studied the influence of DEF on concrete under compressive loads and quantified its crack propagation via CT scan. These works evidently highlight the ability of micro-CT to detect damage caused by various ISRs. Therefore, tomography techniques have immense potential to provide 3D images of ISR-affected concrete without the application of techniques to estimate the 3D representation from 2D petrographic damage features. Yet, the limitations of tomography remain with the ability to resolve small objects, such as the width of cracks below its spatial resolution and the low contrast between the cracks with or without reaction products and the bulk cement paste. Leemann and Münch (2019) observed that the addition of caesium and barite to fresh concrete could help enhance contrast and thus facilitate recognition of ASR secondary products within aggregate particles or cement paste, respectively. Otherwise, contrast remains a challenge for other ISR mechanisms. Nonetheless, further research is still required to improve the current limitations of micro-CT scan, along with enhancing its potential to become a reliable technique to quantitatively appraise induced expansion and damage in concrete caused by ISR.

REFERENCES

Ahmed, H., Zahedi, A., Sanchez, L. F. M., & Fecteau, P.-L. (2022). Condition assessment of ASR-affected reinforced concrete columns after nearly 20 years in service. *Construction & Building Materials, 347*, 128570. https://doi.org/10.1016/j.conbuildmat.2022.128570

ASTM C295. (2012). *Standard guide for petrographic examination of aggregates for concrete, ASTM International, West Conshohocken (USA).*

ASTM C295. (2019). *Standard guide for petrographic examination of aggregates for concrete.*

ASTM C856-20. (2020). Standard practice for petrographic examination of hardened concrete. In *Annual book of ASTM standards.* https://doi.org/10.1520/C0856

ASTM C1723-16. (2016). Standard Guide for Examination of Hardened Concrete Using Scanning Electron Microscopy. In *Annual book of ASTM standards.* https://doi.org/10.1520/C1723-16.can

Bezerra, A. (2020). *The Use of Artificial Intelligence for Assessing Damage in Concrete Affected by Alkali-Silica Reaction (ASR)* [Master's Thesis.]. University of Ottawa.

British Cement Association (BCA). (1992). *The diagnosis of alkali-silica reaction* (p. 44) [Report of a Working Party]. British Cement Association (BCA).

Broekmans, M. (2012). Chapter 7: Deleterious reactions of aggregate with alkalis in concrete. *Reviews in Mineralogy & Geochemistry, 74*, 279–364.

CUR-Recommendation-102. (2008). *Inspection and assessment of concrete structures in which ASR is suspected or has been confirmed, official English translation* (p. 31). Gouda, Netherlands: Centre for Civil Engineering Research and Codes.

Dehghan, A., Zhang, P., Ossetchkina, E., Sloan, D., & Peterson, K. (2016). Digital microscopy applied to damage rating index for alkali-silica reaction in concrete. In D. Cong & D. Broton (Eds.), *Advances in cement analysis and concrete petrography* (pp. 105–125). ASTM International. https://doi.org/10.1520/STP161320180003

Erlin, B., & Stark, D. (1990). Petrography applied to concrete and concrete aggregates. *American Society for Testing and Materials, STP 1061.*

Fernandes, I., Ribeiro, M., Broekmans, M. A. T. M., & Sims, I. (2016). *Petrographic Atlas: Characterisation of Aggregates Regarding Potential Reactivity to Alkalis: RILEM TC 219-ACS Recommended Guidance AAR-1.2, for Use with the RILEM AAR-1.1 Petrographic Examination Method* (1st ed.). Springer Netherlands. https://doi.org/10.1007/978-94-017-7383-6

Fonseca, P. C., & Scherer, G. W. (2015). An image analysis procedure to quantify the air void system of mortar and concrete. *Materials and Structures*, 48(10), 3087–3098. https://doi.org/10.1617/s11527-014-0381-9

Goldstein, J., Newbury, D., Joy, D., Lyman, C., Echlin, P., Lifshin, E., Sawyer, L., & Michael, J. (2003). *Scanning electron microscopy and X-ray microanalysis. A text for biologists, materials scientists, and geologists* (3rd ed.). Berlin: Springer Verlag.

Grattan-Bellew, P. E., & Danay, A. (1992). Comparison of Laboratory and Field Evaluation of Alkali-Silica Reaction in Large Dams. Proc. *International Conference on Concrete Alkali-Aggregate Reactions in Hydroelectric Plants and Dams, Sept 28th to October 2, 1992.* Canadian Electrical Association in Association with Canadian National Committee of the International Commission of Large Dams, pp. 23.

Hernández-Cruz, D., Hargis, C. W., Dominowski, J., Radler, M. J., & Monteiro, P. J. M. (2016). Fiber reinforced mortar affected by alkali-silica reaction: A study by synchrotron microtomography. *Cement and Concrete Composites*, 68, 123–130. https://doi.org/10.1016/j.cemconcomp.2016.02.003

Hollis, N., Walker, D., Lane, S., & Stutzman, P. E. (2006). *Petrographic methods of examining hardened concrete: A petrographic manual, FHWA-HRT-04-150.*

Ingham, J. P. (2013). *Geomaterials under the microscope.* Academic Press.

Jana, D. (2005). Concrete Petrography – Past, Present, and Future. *10th Euroseminar on Microscopy Applied to Building Materials*, Scotland.

Jensen, V., & Sujjavanich, S. (2016). *Alkali silica reaction in concrete foundations in Thailand. 16th International Conference on Alkali-Aggregate Reactions in Concrete (ICAAR)*, Sao Paulo, Brazil.

Joshi, N. R., Matsumoto, A., Asamoto, S., Miura, T., & Kawabata, Y. (2022). Investigation of the mechanical behaviour of concrete with severe delayed ettringite formation expansion focusing on internal damage propagation under various compressive loading patterns. *Cement and Concrete Composites*, 128, 104433. https://doi.org/10.1016/j.cemconcomp.2022.104433

Leemann, A., Katayama, T., Fernandes, I., & Broekmans, M. A. T. M. (2016). Types of alkali-aggregate reactions and the products formed. *Proceedings of the Institution of Civil Engineers - Construction Materials*, 169, 128–135. https://doi.org/10.1680/jcoma.15.00059

Leemann, A., & Münch, B. (2019). The addition of caesium to concrete with alkali-silica reaction: Implications on product identification and recognition of the reaction sequence. *Cement and Concrete Research*, 120, 27–35. https://doi.org/10.1016/j.cemconres.2019.03.016

Lindgård, J., Haugen, M., Castro, N., & Thomas, M. D. A. (2012). *Advantages of using plane polished section analysis as part of microstructural analyses to describe internal cracking due to alkali-silica reactions. 14th International Conference on Alkali-Aggregate Reactions in Concrete (ICAAR)*, Austin, Texas.

Lindgård, J., Skjølsvold, O., Haugen, M., Hagelia, P., & Wigum, B. J. (2004). *Experience from evaluation of degree of damage in fluorescent impregnated plan polished sections of half cores based on the cracking index method. 12th International Conference on Alkali-Aggregate Reactions in Concrete (ICAAR)*, Beijing, China.

Loser, R., & Leemann, A. (2016). An accelerated sulfate resistance test for concrete. *Materials and Structures, 49*(8), 3445–3457. https://doi.org/10.1617/s11527-015-0731-2

Marinoni, N., Voltolini, M., Mancini, L., Vignola, P., Pagani, A., & Pavese, A. (2009). An investigation of mortars affected by alkali-silica reaction by X-ray synchrotron microtomography: A preliminary study. *Journal of Materials Science, 44*(21), 5815–5823. https://doi.org/10.1007/s10853-009-3817-9

Melo, R. H. R. Q., Hasparyk, N. P., & Tiecher, F. (2023). Assessment of concrete impairments over time triggered by DEF. *Journal of Materials in Civil Engineering, 35*(8). https://doi.org/10.1061/JMCEE7.MTENG-15041

Poole, A. B., & Sims, I. (2016). *Concrete petrography, a handbook of investigative techniques.* (2nd ed.). London: CRC Press (Taylor & Francis Group).

Rivard, P. (1998). *Quantification de l'endommagement Du Béton Atteint de Réaction Alcalis-Silice Par Analyse d'images* [Master's Thesis.]. Université de Sherbrooke.

Rivard, P., Fournier, B., & Ballivy, G. (2000). Quantitative petrographic technique for concrete damage due to ASR: Experimental and application. *Cement, Concrete and Aggregates, 22*(1), 63–72. https://doi.org/10.1520/CCA10465J

Sanchez, L. F. M., Drimalas, T., & Fournier, B. (2020). Assessing condition of concrete affected by internal swelling reactions (ISR) through the Damage Rating Index (DRI). *Cement, 1–2*, 100001. https://doi.org/10.1016/j.cement.2020.100001

Sanchez, L. F. M., Drimalas, T., Fournier, B., Mitchell, D., & Bastien, J. (2018). Comprehensive damage assessment in concrete affected by different internal swelling reaction (ISR) mechanisms. *Cement and Concrete Research, 107*, 284–303. https://doi.org/10.1016/j.cemconres.2018.02.017

Sanchez, L. F. M., Fournier, B., Jolin, M., & Duchesne, J. (2015). Reliable quantification of AAR damage through assessment of the Damage Rating Index (DRI). *Cement and Concrete Research, 67*, 74–92. https://doi.org/10.1016/j.cemconres.2014.08.002

Sanchez, L. F. M., Fournier, B., Jolin, M., Mitchell, D., & Bastien, J. (2017). Overall assessment of Alkali-Aggregate Reaction (AAR) in concretes presenting different strengths and incorporating a wide range of reactive aggregate types and natures. *Cement and Concrete Research, 93*, 17–31. https://doi.org/10.1016/j.cemconres.2016.12.001

Shakoorioskooie, M., Griffa, M., Leemann, A., Zboray, R., & Lura, P. (2021). Alkali-silica reaction products and cracks: X-ray micro-tomography-based analysis of their spatial-temporal evolution at a mesoscale. *Cement and Concrete Research, 150*, 106593. https://doi.org/10.1016/j.cemconres.2021.106593

Sims, I., Hunt, B., & Miglio, B. (1992). *Quantifying microscopical examinations of concrete for Alkali Aggregate Reactions (AAR) and other durability aspects.*

Song, Y., Zou, R., Castaneda, D. I., Riding, K. A., & Lange, D. A. (2017). Advances in measuring air-void parameters in hardened concrete using a flatbed scanner. *Journal of Testing and Evaluation, 45*(5), 20150424. https://doi.org/10.1520/JTE20150424

Suzuki, T., Shiotani, T., & Ohtsu, M. (2017). Evaluation of cracking damage in freeze-thawed concrete using acoustic emission and X-ray CT image. *Construction & Building Materials, 136*, 619–626. https://doi.org/10.1016/j.conbuildmat.2016.09.013

Thomas, M., Folliard, K., Drimalas, T., & Ramlochan, T. (2008). Diagnosing delayed ettringite formation in concrete structures. *Cement and Concrete Research*, *38*(6), 841–847. https://doi.org/10.1016/j.cemconres.2008.01.003

Villeneuve, V. (2011). *Détermination de l'endommagement du béton par méthode pétrographique quantitative* [M.Sc.]. Université Laval.

Voltolini, M., Marinoni, N., & Mancini, L. (2011). Synchrotron X-ray computed microtomography investigation of a mortar affected by alkali-silica reaction: A quantitative characterization of its microstructural features. *Journal of Materials Science*, *46*(20), 6633–6641. https://doi.org/10.1007/s10853-011-5614-5

Walker, H. N., Lane, D. S., & Stutzman, Paul E. (2006). *Petrographic methods of examining hardened concrete: A petrographic manual* (FHWA-HRT-04-150; p. 351). Federal Highway Administration (FHWA).

Chapter 6

Mechanical tools

6.1 INTRODUCTION

Well-designed and "sound" concrete generally presents good strength and modulus of elasticity (ME), reasonable tensile strength, a brittle response under uniaxial loading (i.e., compression or tension) and an increase in both ductility and strength (in compression and tension) in a confined environment (Crouch & Wood 1990). This sensitivity to confinement is related to the presence of small flaws or microcracks in the material. Thus, even for "sound" concrete under a triaxial compression load, there will always be locations in tension within the bulk material due to its heterogeneous and "defective" nature (Crouch & Wood 1990). Moreover, the complexity of concrete behaviour under stress is even greater when the material is deteriorated (Crouch & Wood 1990). Hence, the understanding of the mechanical properties changes (i.e., compressive and tensile strengths, ME, aggregate interlock through the direct shear, and also the stress/strain relationship) as a function of ISR-induced development is a very important step for assessing condition (i.e., damage extent), along with designing rehabilitation strategies for affected structures and structural members (Kubo & Nakata 2012).

Different ISR mechanisms may influence differently the mechanical properties of affected concrete (Sanchez et al., 2018). This dissimilar impact is related to the unique microscopic distress features and general "pattern" associated with each of these deterioration mechanisms (see Chapter 5; Sanchez et al. 2020). To understand the impact of the various ISRs on the mechanical response of affected concrete, it is crucial to first discuss the response of conventional and "sound" concrete with respect to standardized and advanced mechanical test procedures conducted in compression, tension and shear. Depending on the "sound" concrete response and failure mode, a given mechanical test procedure may or may not be considered suitable or "diagnostic" to appraise the damage extent of ISR-affected concrete.

DOI: 10.1201/9781003188155-6

6.2 TEST PROCEDURES IN COMPRESSION

6.2.1 Compressive strength test

Compressive strength is the most common mechanical test procedure performed in concrete due to four main reasons: (1) it is assumed that most of the mechanical properties of concrete are directly related to compressive strength (although not always true); (2) concrete is primarily used in compression due to its low tensile strength; (3) structural design codes are mainly based on the compressive strength; (4) the compressive strength test is a fast, easy and relatively inexpensive procedure (Mindess et al. 2003). This method is commonly implemented as per national/international standards such as ASTM C39 or similar (ASTM C39 2003). The compressive strength test outcomes may vary as a function of a number of parameters such as temperature and humidity, sample size and geometry, loading rate, etc. Therefore, it is important that the aforementioned parameters be properly controlled so that the test outcomes gathered from different laboratories can be compared to one another. In this section, some of the most important parameters of the test will be discussed; further information can be found in (Mindess et al. 2003; ASTM C39 2003).

Compressive strength tests as per ASTM C 39 are conducted using cylindrical specimens or cores with a length-to-diameter ratio (l/d) of 2:1 (i.e., usually 100 mm by 200 mm or 150 mm by 300 mm specimens). If the specimens are manufactured in the laboratory, they should be moulded in layers, depending on the method of consolidation; if the specimens are rodded, three equal layers are required. However, if they are vibrated, only two layers are to be used. Once manufactured, the specimens should be moist cured in a standard moist room or in saturated lime water at 23 ± 2°C until testing (normally over 28 days or as per the project's rationale). Conversely, if the test is conducted on cores extracted from concrete structures, variations in the moisture condition of the specimens from coring to testing may occur, which can significantly influence the test results (ASTM C42 2008). As per ASTM C 42, cores should be tested "in the same moisture condition than that they were in the field". However, since moisture gradients are often observed within concrete elements (in contact or not with external moisture sources), it is recommended that upon coring, the specimens should be wrapped in an impermeable plastic film and stored for at least five days before testing (ASTM C42 2008). Moreover, according to Canadian Standards, cores extracted from structures should be rewetted for 48h in a standard moist curing room prior to testing in order to reset their moisture conditions (CSA.A23.2-14C 2009).

After removal from the moist room and before testing, the specimen's ends should be treated to become smooth, plane and perpendicular to the longitudinal axis. Normally, planeness is achieved by mechanical grinding or by capping through the use of stiff Portland cement pastes, high-strength gypsum plasters or even sulphur mortars. Once the ends are smooth and flat, the specimen is ready for testing. Loading is then applied to the specimen at a rate ranging from 0.15 to 0.35 Mpa/s for hydraulic machines or at a deformation rate of 1 mm/min for mechanical machines until failure. The maximum load and type of failure are finally reported. It is worth noting that different geometries (such as cubes) and test conditions are used in other standards and countries, particularly in some European countries (Mindess et al. 2003). Correction factors should be then implemented to compare the test outcomes bearing distinct parameters. Table 6.1 illustrates the correction factors normally adopted to correlate compressive strength responses of cylinders and cubes.

Regardless of the test conditions, the response of concrete under non-confined, uniaxial compression is considered a progressive phenomenon related to the generation and propagation of microcracks in the system (Mindess et al. 2003). At the macroscale, cracks are generated parallel to the loading direction (i.e., vertically), forming a conic shape at the top and bottom ends of the specimen due to the existing friction between the ends and the loading plate. In a condition of non-friction, completely parallel and vertically aligned cracks would be expected before failure. Figure 6.1 illustrates the macroscale cracking process. Disregarding the presence of friction, an important number of vertical cracks is nevertheless generated; these vertical cracks create various "slender little columns" in the specimen, which leads to local buckling. Ultimately, failure in unconfined uniaxial compression is caused by tension failure of multiple and simultaneous planes.

On the other hand, at the micro-mesoscale, the phenomenon of failure under a non-confined uniaxial compression is slightly more progressive. First, cracks are initially generated at the interface between the aggregate

Table 6.1 Correction factors used to correlate compressive strength results from cylinders and cubes

Cube strength (Mpa)	Cylinder strength (Mpa)	Correction factor
9	7	1.29
16	12	1.33
20	15.5	1.29
24.5	20	1.23
27.5	23	1.20
35.5	32	1.11
42	36	1.17
44	37.5	1.17
48	44	1.09
52	50.5	1.03

Source: Adapted from (Mindess et al. 2003)

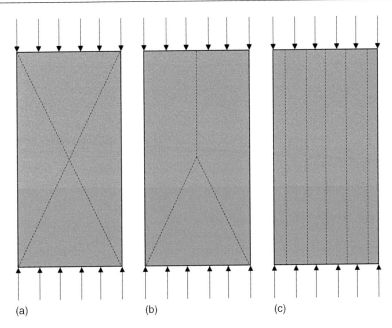

Figure 6.1 Macroscale cracking formation in an unconfined uniaxial compressive strength test: (a) confinement at both ends, (b) confinement at the bottom and splitting failure at the top and (c) splitting failure.

(Adapted from Mindess et al. 2003; Swamy, 1979.)

particles and the cement paste, the so-called interfacial transition zone (ITZ), when the load reaches about 30%–40% of the ultimate capacity of the material being tested; then, a process of "slow/stable crack propagation" takes place where the tensile strength of the material is locally reached due to stress intensity factors and thus, the load is transferred elsewhere (Mindess et al. 2003). The propagation of cracks generated at the ITZ slowly continues with loading, and at 50%–70% of the ultimate capacity, multiple cracks reach the cement paste. Some of those cracks are "arrested" by aggregate particles present in the granular system and need to outline them to keep propagating, while others remain free in the cement paste. At about 75% of the ultimate capacity, these cracks increase significantly in length and width, and some of them start linking to one another. If the load is sustained, failure may take place from this stage and onwards. At 95%–100% of the ultimate capacity, most of the cracks are linked, and the phenomenon becomes "unstable", with fast crack propagation leading to failure (Mindess et al. 2003). Figure 6.2 illustrates the failure process previously described.

Understanding the failure process of "sound" concrete in unconfined uniaxial compression is imperative to evaluate whether this test procedure may be considered a diagnostic tool to appraise ISR-deteriorated concrete. The literature suggests that the compressive strength of ISR-affected concrete

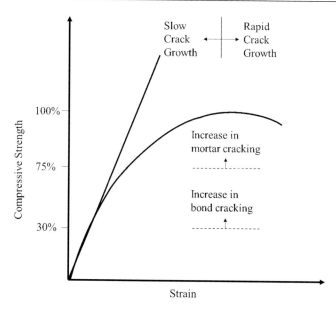

Figure 6.2 Concrete failure in unconfined uniaxial compression.
(Adapted from Mindess et al. 2003.)

varies as a function of the damage mechanism. Figure 6.3 illustrates the impact of different ISRs on the compressive strength of affected concrete. For alkali-aggregate reaction (AAR)–affected concrete, the compressive strength decrease is only significant (i.e., 15% or more) at high and very high expansion levels (i.e., > 0.20% of expansion or strain; Nixon & Bollinghaus 1985; Pleau et al. 1989; Smaoui, Bérubé, et al., 2004b; Sanchez et al., 2018, 2017). This happens particularly because at low and moderate expansion levels (i.e., 0.05%–0.12% of expansion or strain), the vast majority of AAR-induced cracks are generated within the aggregate particles (Sanchez et al., 2015a; Sanchez, Fournier, Jolin, Bedoya, et al., 2016); in other words, the AAR-induced cracks do not directly interact with the cracks formed over the loading process in compression, as previously described in this section. Otherwise, the response of concrete affected by other ISR mechanisms, for example, delayed ettringite formation (DEF) and freeze and thaw (FT), is slightly different, especially at the early stages of the reaction. DEF- and FT-affected concrete display a higher compressive strength reduction (i.e., 10% to 20%) for low and moderate expansion levels (i.e., 0.05%–0.12% of expansion or strain) than AAR-affected concrete; this happens since DEF and FT generate microcracks in the cement paste, particularly at the ITZ and bulk cement paste/pores, respectively. Therefore, these cracks interact more with the new cracks developed over the loading process in compression or even increase (i.e., in length and width) while loading, which explains the more pronounced impact. This impact could be

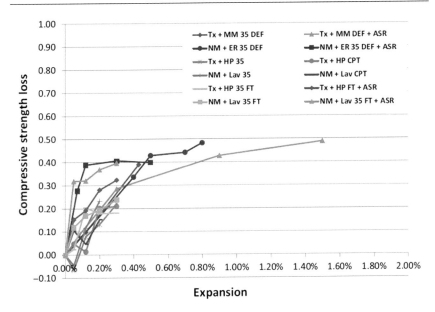

Figure 6.3 Impact of distinct ISR on the compressive strength of concrete. The mixtures presented in the labels are 35 MPa concrete mixtures incorporating either New Mexico or Texas sand as coarse or fine reactive aggregates, respectively. These mixtures developed DEF, FT and ASR deterioration (Sanchez et al., 2018).

even higher in the compressive strength reduction of DEF- and FT-affected concrete; however, two mechanisms are verified to inhibit further losses, such as (1) the cracks being extremely localized, especially in DEF cases, and (2) the "arrest mechanism" provided by the aggregates, which stops the cracks' propagation once they get in contact with the aggregate particles, as described by Mindess et al. (2003)). Conversely, at high and very high expansion levels (i.e., 0.20%–0.30%), the compressive strength reductions are quite similar whatever the ISR mechanism (i.e., 20%–25%) since the cracks generated by the three mechanisms are observed in the cement paste at later stages, which makes a similar interaction with the compression load. Interestingly, combined mechanisms display a more important impact on the compressive strength of affected concrete, particularly at the early stages of DEF + ASR- and FT + ASR-affected concrete, as a few examples; this influence is more pronounced when ASR is triggered from a reactive coarse aggregate. This behaviour indicates a higher interaction of the cracks in the cement paste and aggregates over the compression loading, besides suggesting the decrease of the so-called arrest mechanism previously described since the cracks go through and split the coarse aggregate particles.

The results previously discussed demonstrate that the compressive strength test is not always effective in diagnosing the damage degree (i.e., expansion level) of concrete affected by ISR. On the one hand, it is very important to acknowledge the actual and residual strength of ISR-affected concrete to understand the potential structural implications of the current deterioration in affected structures or structural components. On the other hand, having a test procedure suitable to estimate the current induced expansion of distinct ISR mechanisms is crucial to recognize the potential for further deterioration of affected concrete members. All of the aforementioned should be carefully evaluated prior to the use of compressive strength in practice.

6.2.2 Modulus of elasticity (ME)

The stiffness of concrete is conventionally represented by the ME. However, it is important to notice that concrete is a nonlinear inelastic material under both tension and compression loads. Therefore, the concept of ME needs to be carefully applied in this context since it does not represent a single value as for linear elastic materials (Mindess et al. 2003).

It is widely known and accepted that the stiffness response of various materials, such as aggregates, cement paste, mortar and concrete, is dissimilar (Kosmatka et al. 2003). Figure 6.4 illustrates the stress-strain relationship of the previously mentioned materials. Normally, aggregates display higher stiffness than cementitious composites (i.e., cement paste, mortar and concrete), although the stiffness response of aggregates depends on their nature (i.e., lithotype). Moreover, an almost elastic linear behaviour is observed in the stress-strain curve of aggregates' specimens. Conversely, one

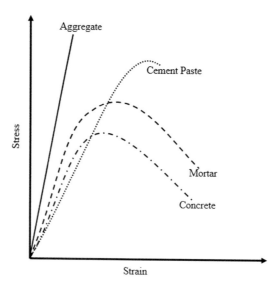

Figure 6.4 Stress vs strain relationship for various materials.

(Adapted from Kosmatka et al. 2003.)

might expect that the cement paste (which does not bear aggregates) is the material displaying the lowest stiffness, along with nonlinear inelastic behaviour. Concrete and mortar mixtures are within these two extremes (Kosmatka et al. 2003) due to the inclusion of aggregates, which display a higher stiffness.

There are several possibilities for adopting the ME of concrete. Figure 6.5 illustrates some of the moduli that can be gathered from the stress-strain curve of concrete in compression. The initial tangent modulus is likely the closest approximation of the ME derived from a linear elastic material. This parameter is not conventionally used in the design of concrete structures since it is considered to overestimate the stiffness of the material, besides being captured with small values of stresses and strains that do not match the ones applied in practice. Otherwise, the secant ME (i.e., the slope of the secant between the origin and a selected point in the stress-strain curve) bears an element of nonlinearity, being a more conservative parameter than the initial tangent modulus and better represents the behaviour of the material in the field (Mindess et al. 2003). The secant ME (normally the slope between the origin and 40% of the ultimate stress) is often adopted to design concrete structures. The initial tangent or the secant moduli are not always easily determined due to errors in measurement caused by the specimen's seating at the beginning of the test or by the presence of cracks in the specimen's evaluated that close under load (Mindess et al. 2003). As such, the chord modulus, as per ASTM C 469, is used, which is the slope of a line drawn between two points on the stress-strain curve. The initial tangent modulus typically corresponds to the slope of the curve at a strain of

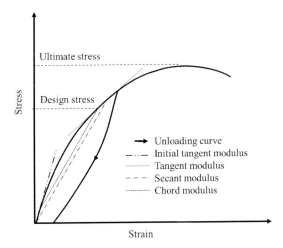

Figure 6.5 Stress-strain curve of concrete in compression displaying the various elastic moduli.

(Adapted from Mindess et al. 2003.)

0.00005 (lower limit) and 40% of the compressive strength value (upper limit) of the material tested (ASTM C 469 2006). The chord modulus is a more conservative measure than the initial tangent modulus and is easily gathered experimentally (Mindess et al. 2003).

The specimens (i.e., laboratory cylinders or cores) should be manufactured or cored, moisture conditioned and their ends treated similarly to the compressive strength test to perform the ME test as per ASTM C 469. Then, a load of 0.25 MPa/s ± 0.05 MPa/s at a constant rate should be applied until it reaches 40% of the compressive strength of the concrete under analysis. Upon reaching this value, the load is then reduced to zero at the same rate. Three loading-unloading cycles are normally performed, where the first cycle is discarded. The average of the second and third cycles is often selected to represent the stiffness or ME of the concrete (ASTM C 469 2006).

Understanding the stiffness response of concrete, normally adopted as the secant or chord ME, is crucial to appraise its efficiency as a diagnostic protocol to assess ISR-affected concrete. As for the compressive strength, the literature suggests that the ME of ISR-affected concrete varies as a function of the damage mechanism. Figure 6.6 illustrates the impact of different ISRs on the ME of affected concrete. For AAR-affected concrete, the ME is already significantly affected at low and moderate expansion levels (i.e., 0.05%–0.12% of expansion or strain), where reductions of until 30% are

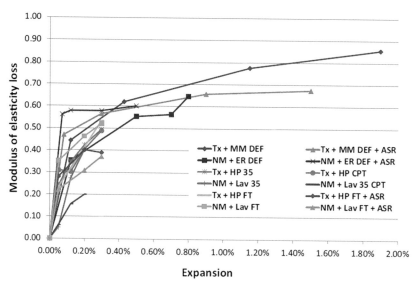

Figure 6.6 Impact of distinct ISR on the ME of affected concrete. The mixtures presented in the labels are 35 Mpa concrete mixtures incorporating either NM or TX as coarse or fine reactive aggregates, respectively. These mixtures developed DEF, FT and ASR deterioration (Sanchez et al., 2018).

frequently observed (Sanchez et al., 2017, 2018). This happens due to the generation of an important number of cracks within the reactive aggregate particles (i.e., coarse and or fine; Sanchez, Fournier, Jolin, Bedoya, et al., 2016; Sanchez et al. 2015a), thus weakening the stiffness of the aggregate. As the induced expansion keeps progressing, the cracks previously formed in the aggregates expand to the cement paste. Therefore, the ME of the affected concrete keeps decreasing but at a lower degree. For very high expansion levels (i.e., 0.30% of expansion or strain), the ME reduction of AAR-affected concrete can reach 50% (Sanchez et al., 2018, 2017). Otherwise, the response of concrete affected by other ISR mechanisms, for example, DEF and FT, is different, especially at the early stages of the damage process. DEF- and FT-affected concrete display a higher decrease in ME at low and moderate expansion levels (i.e., 0.05%–0.12%) than AAR-affected concrete; this happens because DEF and FT generate an important number of microcracks at the interface between aggregates and cement paste (i.e., ITZ). Therefore, these cracks may "break" the bond between the aggregates and cement paste and thus significantly drop the stiffness of the material as the stress transfer "bridge" created by that bond is eliminated. The decrease in ME of DEF- and FT-affected concrete may vary from either 35% to 55% for DEF cases or 25% to 40% for FT cases at low and moderate expansion levels (Sanchez et al., 2018). Conversely, at high and very high expansion levels (i.e., 0.20%–0.30% of expansion or strain), the ME reductions are quite similar, whatever the ISR mechanism (i.e., 50%–60%), since the cracks generated by the three mechanisms are very spread in the affected concrete at later stages (i.e., cement paste, ITZ and aggregates), which makes a roughly similar interaction with a controlled compression load. Interestingly, combined mechanisms display a dissimilar impact on the ME of affected concrete. DEF + ASR coupling seems to further reduce the ME of the affected concrete, whereas FT + ASR coupling seems to lessen the impact on the stiffness of the material; the more pronounced influence of DEF + ASR coupling is likely due to the higher chemical affinity and microscopic distress features interaction (i.e., sharper and wider cracks at the ITZ and aggregate particles) present in this combined mechanism. Nevertheless, the results discussed in this section demonstrate that the ME is a suitable mechanical procedure to appraise the current stage of ISR-affected concrete.

6.2.3 Stiffness damage test (SDT)

Concrete under cyclic loading behaves quite differently than under sustained loading. During repeated cycles in compression, concrete displays a change in the stress-strain curve from a concave downward to a concave upward shape (Mindess et al. 2003). This change is also characterized by an increase in secant modulus (i.e., stiffness) over the initial cycles followed by a reduction of stiffness or secant modulus at later stages. Furthermore, the accumulation of strain is rapid during the initial cycles, slows for the intermediate cycles and increases again for the final cycles prior to failure. Figure 6.7 illustrates the stress-strain curves of concrete under cyclic load in compression.

The shape of the stress-strain curves at early cycles suggests that pre-existing defects and or microcracks are closed over the first compression cycles and may be reopened and propagated at later cyclic stages. The mechanism of damage development and failure under cyclic loading is explained through fracture mechanics concepts (i.e., stress concentration), where important stresses are generated at the tip of the flaws, defects and microcracks generated during the hydration process and often observed at the ITZ of concrete (Mindess et al. 2003). At the microscale, when concrete is tested under repeated cycles of loading, these pre-existing flaws and cracks tend to grow. This is demonstrated in Figure 6.7 by the formation of hysteresis loops. The hysteresis area loops represent the irreversible energy of deformation, some of which represent the sliding across surfaces of cracks, while others demonstrate the energy to extend the pre-existing cracks (Mindess et al. 2003). The hysteresis area loops decrease at first with successive load cycles but then begin to increase again prior to failure. In the beginning, the crack growth is slow, yet due to stress concentrations, pre-existing microcracks eventually increase in size (i.e., length and width) with the rise in the number of cycles and interactions with the aggregates in the cementitious system, which leads to the stabilization of larger cracks growth; however, the energy supplied by the repeated loadings eventually increases the deterioration at the cracks tips, leading to crack propagation until failure (Mindess et al. 2003).

If one evaluates the mechanism of failure under cyclic loading, along with the changes in the distinct variables that comprise the test outcomes, such as

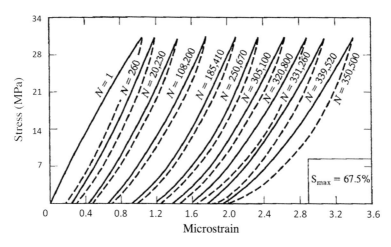

Figure 6.7 Stress-strain curves of concrete under cyclic loading.

(Adapted from Mindess et al. 2003.)

the stress-strain curve shapes, the hysteresis area loops and plastic deformation (or strain), one verifies an opportunity of using cyclic tests for assessing pre-existing deterioration in concrete. In this context, the SDT has been proposed, and, differently from the compressive strength and ME, it is a fairly new procedure; its use has recently been adopted to assess the condition of concrete affected by ISR; hence, a detailed description of the test development is deemed appropriate herein.

The SDT is a mechanical and cyclic test procedure used to assess the condition (i.e., damage extent) of concrete affected by ISR mechanisms (Sanchez, Fournier, et al., 2014). The SDT was initially developed by Walsh, who observed a good correlation between the crack density and the cycles of loading/unloading (i.e., stress/strain relationship) of rock specimens (Walsh 1965); Crouch then adapted this procedure for concrete specimens in 1987 (Crouch 1987). Following those developments, Chrisp et al. proposed the use of the SDT for assessing concrete affected by ASR through the application of a fixed stress of 5.5 MPa at a loading rate of 0.10 MPa/s (Chrisp et al. 1993; Chrisp et al. 1989); the authors initially wanted the procedure to remain "non-destructive", thus enabling the reuse of the test specimens for further analyses. Chrisp and coworkers conducted over 1,000 tests on cores extracted from ASR-affected concrete structures; after appraising the stress-strain curves of the deteriorated specimens, the authors proposed the following test outcomes to assess the damage extent of affected concrete (Chrisp et al. 1993; Chrisp et al. 1989):

- ME (E): average secant ME of the last four cycles since affected concrete often demonstrates lower ME than undamaged or sound specimens.
- Hysteresis area (H, in J/m^3): area of the hysteresis loops (i.e., area under the stress-strain curves) averaged over the last four cycles since affected concrete displays higher energy loss (or hysteresis area loops) than undamaged or sound concrete.
- Nonlinearity index (NLI): it represents the ratio between the slope of the stress-strain curve at half of the maximum load and the secant ME (E). This parameter provides information on the damage extent and pattern (i.e., cracks orientation).

Chrisp et al. observed that the hysteresis area of the first cycle was much greater than that of the following four consecutive cycles; this behaviour has been attributed to the sliding effect across open crack surfaces and their closure upon loading. Therefore, the results from the first cycle were suggested to be disregarded to only obtain a response from the material itself. Moreover, the ME was found to be the most sensitive test outcome for slightly damaged concrete. However, for higher degrees of deterioration, the hysteresis area has been verified as the critical test outcome for assessing the extent of damage. Finally, it has been observed that the crack pattern could influence the results of the test since deteriorated concrete presenting cracks mostly oriented perpendicular to the load direction were found to display a low ME, a larger hysteresis area and an NLI greater than unity, while those with cracks mainly oriented parallel to the load direction were seen to yield a high ME, a smaller hysteresis area and an NLI lower than unity (Chrisp et al. 1993; Chrisp et al. 1989).

Later, Smaoui et al. (Smaoui, Fournier et al., 2004a) further evaluated the efficiency of the SDT on laboratory-made concrete specimens mix-proportioned as per the concrete prism test (CPT) according to ASTM C 1293, incorporating a variety of reactive aggregate types (i.e., fine versus coarse) and natures (i.e., lithotypes), and presenting various expansion levels following storage at 38°C at 100% R.H. (ASTM C1293 2015). After performing numerous tests, it has been verified that the most efficient SDT outcome was the hysteresis area of the first cycle; moreover, they observed that the SDT should be conducted with a fixed load of 10 MPa instead of 5.5 MPa to increase the diagnostic character of the test since lower stress levels were found to be unsuitable to allow ASR-induced microcracks to sufficiently reclose, which decreased the test efficiency and reliability. Smaoui et al. also verified that the correlation between the expansion and the plastic deformation obtained after the five loading/unloading cycles was fairly satisfactory. However, they noted a high variability for both the hysteresis area and plastic deformation of ASR-affected concrete incorporating different reactive aggregates. These variabilities were found to be possibly associated with the type (fine or coarse) and nature (lithotypes) of reactive aggregates used in concrete, which could lead to distinct deterioration patterns (i.e., cracks density, orientation, locations, etc.; Smaoui, Fournier et al., 2004a).

Sanchez et al. (Sanchez, Fournier, et al., 2014, Sanchez et al. 2015b, Sanchez, Fournier, Jolin, Bastien, et al. 2016) performed a comprehensive experimental campaign, evaluating numerous concrete mixtures presenting distinct mechanical properties and reactive aggregates, on the use of SDT for appraising concrete affected by AAR but also deteriorated by other ISR mechanisms such as DEF and FT, single or combined. Further details on the procedure developments and specific considerations may be found in Sanchez et al. (Sanchez, Fournier, et al. 2014, Sanchez et al. 2015b, Sanchez, Fournier, Jolin, Bastien, et al. 2016). Briefly, it has been found that the SDT should be conducted at 40% of the material's mechanical capacity at the loading rate proposed by Chrisp et al. of 0.10 MPa/s. Furthermore, the method was considered statistically efficient and reliable for assessing the damage extent of ISR-affected concrete, especially through the use of the Stiffness Damage Index (SDI) and Plastic Deformation Index (PDI) outcomes; the SDI and PDI represent, respectively, the ratio of dissipated energy/plastic deformation to the total energy/deformation implemented in the system over five cycles of loading-unloading (i.e., SI/(SI + SII) and DI/(DI + DII)). Finally, as per Chrisp et al. (Chrisp et al. 1993; Chrisp et al. 1989), the Nonlinearity Index (NLI), defined as the secant modulus at half the maximum load (Secant 2) divided by the secant modulus at the maximum load (Secant 1) has been confirmed as an interesting outcome to assess the damage extent and pattern in concrete affected by ISR. Figure 6.8 illustrates the SDT set-up and the calculation of the SDI, PDI and NLI parameters.

Besides optimizing the testing load and outcomes, Sanchez et al. (2015b; Sanchez, Fournier, Jolin, Bastien, et al. 2016) also studied some practical parameters that might interfere with the SDT results, such as the specimen's size (i.e., 100 vs 150 mm diameter, distinct length-to-diameter ratios), ends rectification type (i.e., grinding versus capping), moisture condition, drying and rewetting effects, and location of the core in the affected structure (i.e., surface vs core of the members under analysis). It has been verified that the sample's diameter and the end rectification type do not significantly influence the SDT outcomes. However, the sample length-to-diameter ratio, the moisture condition and the specimen's location were indeed found to impact the test results. Therefore, for practical purposes and following the Canadian Standards (CSA.A23.2-14C 2009), it has been proposed that cores extracted from structures should be cut and rectified to have a length-to-diameter ratio close to 2:1; moreover, the specimens should be rewetted for 48 h in a standard moist curing room prior to stiffness damage testing to decrease the test variability. Figure 6.9 illustrates a flowchart developed by Sanchez et al. to describe the practical tasks required before conducting the SDT in the laboratory (Sanchez et al. 2015b; Sanchez, Fournier, Jolin, Bastien, et al., 2016).

Most of the works developed so far using the SDT to assess the damage extent of affected concrete were conducted evaluating AAR-affected concrete. However, the test procedure has been shown to also be effective and reliable in appraising other ISR mechanisms such as DEF and FT. Moreover,

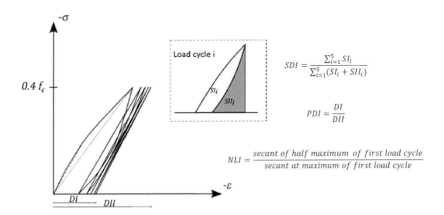

$$SDI = \frac{\sum_{i=1}^{5} SI_i}{\sum_{i=1}^{5}(SI_i + SII_i)}$$

$$PDI = \frac{DI}{DII}$$

$$NLI = \frac{secant\ of\ half\ maximum\ of\ first\ load\ cycle}{secant\ at\ maximum\ of\ first\ load\ cycle}$$

Figure 6.8 SDT set-up and stress-strain curve of ISR-affected concrete: calculation of the indices SDI, PDI and NLI.

(Adapted from Kongshaug et al., 2002.)

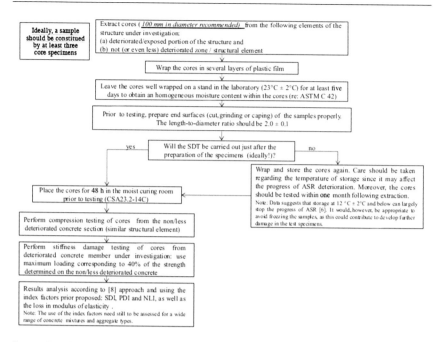

Figure 6.9 SDT flowchart displaying practical tasks required to conduct the test in the laboratory.

(Adapted from Sanchez et al. 2015b; Sanchez, Fournier, Jolin, Bastien, et al. 2016.)

as previously discussed in this section, the SDT outcomes (i.e., particularly the SDI and PDI) measure the number of inner cracks (or physical integrity) of affected concrete. Therefore, these parameters should not largely vary from one ISR mechanism to another as a function of induced development. Figure 6.10 demonstrates the SDI and PDI values of concrete mixtures affected by ASR, DEF and FT (Sanchez et al. 2018). Analysing the data, it is clear that both SDI and PDI are very similar for ASR and FT mechanisms, varying from 0.10 to 0.45 and from 0.08 to 0.45, respectively, for low (i.e., 0.05%) and very high (i.e., 0.30%) expansion levels. Conversely, the results obtained for DEF, although following the same trend, are slightly lower than the ones for ASR and FT, ranging from 0.10 to 0.25 and from 0.08 to 0.20, respectively. These findings might be related to (1) the thinner and more spread crack pattern found in ASR and FT deterioration when compared to the sharper and more localized pattern observed in DEF-affected concrete and (2) the constantly observed cracks outlining the aggregate particles (i.e., ITZ) in DEF-affected concrete, which could limit their full closure under compressive loads (Sanchez et al. 2018). It should be noted that SDI and PDI values under 0.10 generally represent sound conventional concrete.

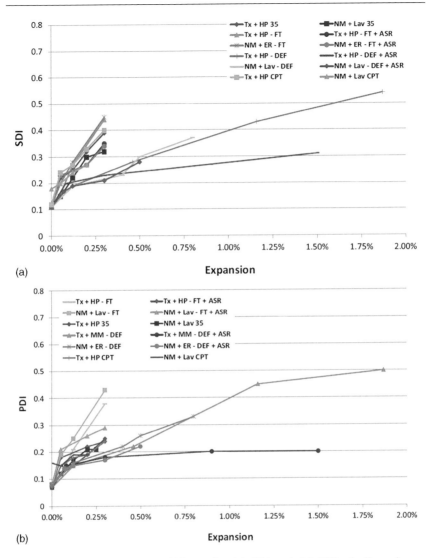

Figure 6.10 Impact of distinct ISR on the (a) SDI and (b) PDI of affected concrete. The mixtures presented in the labels are 35 MPa concrete
mixtures incorporating either NM or TX as coarse or fine reactive
aggregates, respectively. These mixtures developed DEF, FT and ASR
deterioration (Sanchez et al., 2018).

The previous results indicate that the SDT is an efficient tool for assessing
the induced expansion and deterioration in concrete caused by distinct ISR
mechanisms. Further studies should be conducted to understand the parameters influencing the SDI and PDI fluctuations for distinct ISR mechanisms,
yet, in general, a consistent deterioration trend is observed for all mechanisms evaluated so far.

6.3 TEST PROCEDURES IN TENSION

6.3.1 Tensile strength test

Tensile strength is a parameter that is not commonly used for designing reinforced concrete structures (other than pavements), and it is usually adopted (for sound and conventional concrete) as being about 10% of the compressive strength of the material at a given age (Mindess et al. 2003; Swamy 1979). Otherwise, the literature suggests that this parameter is much more influenced by ISR deterioration and cracking in general than the compressive strength, and thus its evaluation could likely add interesting information on ISR-induced development (Sanchez, Multon et al., 2014; Sanchez et al., 2017, 2018).

Differently from the compressive strength, the response of "sound" concrete under tension is very brittle and is governed by fracture mechanics principles (Broek 1982; Mindess et al. 2003; Swamy 1979). According to fracture mechanics, the mechanical response of materials containing flaws, defects and microcracks, such as concrete, is very diminished when compared to its theoretical potential. Results very often tend to be between 10 to 100 times lower than theoretically expected (Broek 1982; Mindess et al. 2003; Swamy 1979). Failure takes place as per the formation of macrocracks generated from pre-existing defects and flaws in the material; these macrocracks are formed due to the presence of stress intensity factors at the tips of pre-existing microdefects, increasing their size (i.e., length and width) until they reach the so-called critical length; defects displaying sizes above and beyond the critical length lead to brittle failure of concrete.

At the macroscale, the response of "sound" concrete under tension may be divided into two stages: before and after the formation of the lead macrocrack bearing the critical length. Before such a macrocrack is formed, the response of the material depends on its microstructure quality (i.e., cement paste, ITZ and aggregates; Swamy 1979). Conversely, once the lead macrocrack is formed, the mechanical response is solely dependent on the friction (i.e., slide across surfaces) of the lead macrocrack. On the other hand, at the micro-mesoscale, the phenomenon of failure in tension of "sound" concrete is normally divided into five steps. First, microcracks are developed (normally in the cement paste) from pre-existing flaws or defects in the material at low loading levels (step 1), followed by a process of slow crack propagation (step 2). Then, some of the formed cracks are "arrested" by the aggregate particles present in the granular system and need to outline them to keep propagating, while others remain free in the cement paste (step 3). Finally, the critical crack length is reached (step 4), and thus the process changes from a stable (i.e., slow crack propagation) to an unstable (i.e., fast crack propagation) process, leading to a sudden and brittle failure (step 5) (Swamy 1979). Figure 6.11 illustrates the previously described failure process.

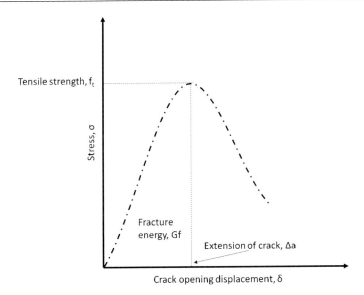

Figure 6.11 Concrete failure in tension.

(Adapted from Swamy 1979.)

Several test set-ups have been developed to assess the tensile strength of concrete; the most common approaches are the direct tension test, the bending tensile test and the splitting tensile test (Mindess et al. 2003). The direct tension test is theoretically a quite suitable technique to evaluate the "pure" or "true" tensile strength of concrete. However, premature failure at the ends of the specimens caused by secondary stresses induced by the grips is quite often verified (Mier & Van Vliet 2002). Therefore, the test has not been adopted by ASTM as a standardized test procedure. Figure 6.12 illustrates the direct tension test set-up normally used. Nevertheless, RILEM displays a recommendation for a direct tensile strength test in concrete that was primarily developed for research (Zheng et al. 2001; Mindess et al. 2003). This method involves applying direct tension to cylindrical or prismatic specimens through end plates glued to the specimens. The ends of the specimens should be rectified and carefully cleaned so that they become flat and smooth, enabling proper adhesion of the glue (normally a polyepoxy resin); they must be perpendicular to the axis of the specimen within 1/4° (Zheng et al. 2001; Mindess et al. 2003). The load is applied at a rate of 0.05MPa/s until failure takes place. The US Bureau of Reclamation also specifies a direct tension test that uses bonded end plates (Zheng et al. 2001; Mindess et al. 2003). Otherwise, the bending tensile and splitting tensile tests are both standardized techniques that measure "indirectly" the tensile strength of concrete; therefore, they are perceived as "less diagnostic" tools to assess ISR-induced damage development since failure happens in designated areas (i.e., failure planes) in both procedures.

Figure 6.12 Direct tension test set-up.

The splitting tensile test as per ASTM C 496 is performed on standard cylinders, tested on their side in diametral compression, as illustrated by Figure 6.13 (ASTM C496 2008). The load is normally applied through a narrow bearing strip of relatively soft material. The tensile stress distribution along the vertical diameter of the specimen is not uniform; there are

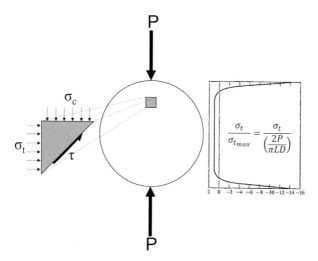

Figure 6.13 Splitting tensile strength test set-up and corresponding stress distribution.

(Adapted from Mindess et al. 2003.)

high compressive stresses near the ends of the vertical diameter while a nearly uniform tensile stress area acting over the middle-two thirds of the specimen takes place (ASTM C496 2008; Mindess et al. 2003). Since concrete is much weaker in tension than in compression, it will split in tension at a much lower load than is required to fail the specimen in compression.

There is no simple relationship between the values of tensile strength found from the splitting test and those measured in direct tension. However, it is commonly assumed that the splitting tensile strength values are 5%–12% higher than the direct strength values, although some works have shown close results from these two test procedures (Zheng et al. 2001; Mindess et al. 2003). Detailed analyses from the splitting tensile strength test have shown that the test as per ASTM C 496 is not a "true" property of the material since the results are dependent on the size, width and type of the load-bearing strips (Zheng et al. 2001; Mindess et al. 2003).

The flexural strength of concrete may be determined as per ASTM C78. Prismatic specimens (i.e., 150 mm by 150 mm by 500 mm) are manufactured in two consecutive and equal layers; each layer is compacted by either rodding (i.e., 60 times per layer) or vibration (as per compression tests). The specimens are cured in a standard fashion and then tested in flexure in four or third-point loading, as illustrated in Figure 6.14 (ASTM C78 2010). The flexural strength of concrete, as per ASTM C78, can also be conducted with specimens sawn from concrete components.

The specimens manufactured in the laboratory should be tested and turned on their sides with respect to their position as moulded. This should provide smooth, plane and parallel faces for loading (ASTM C78 2010; Mindess et al. 2003). Otherwise, the specimens should be ground or capped. The specimens are loaded at a rate of 860–1,200 KPa/min, and the theoretical maximum tensile strength, also called modulus of rupture (R), is calculated according to Equation 6.1.

$$R = Pl / (bd^2), \tag{6.1}$$

Figure 6.14 Flexural test as per ASTM C 78.

(Adapted from Mindess et al. 2003.)

where P is the peak load, l is the span length, b is the specimen width and d is the specimen depth. Equation 6.1 holds if the specimen breaks between the two interior loading points (i.e., the middle third of the beam). If the beam breaks outside of these points, but not more than 5% of the span length, Equation 6.2 should be used (ASTM C78 2010; Mindess et al. 2003).

$$R = 3Pa / (bd^2), \tag{6.2}$$

where a is the average distance between the point of fracture and the nearest support.

The flexural test tends to overestimate the "pure" or "true" tensile strength of concrete by about 50%, especially due to the fact that the flexure formula assumes that the stress varies linearly across the cross-section of the specimen, which is not true due to the nonlinear stress-strain behaviour of concrete. Nevertheless, the flexure test remains quite useful, especially to measure the response of concrete mixtures used in components under bending in the field, such as pavements (Mindess et al. 2003).

The gas pressure tension test, also known as the indirect tension test, is a non-standardized technique that has been developed by the Building Research Council of Waterford (United Kingdom) as a means of investigating anisotropic behaviour of materials along with overcoming the challenges of premature failure of the direct tensile strength. The test procedure uses compressed gas to apply a uniformly distributed pressure to the curved surface of standard (i.e., 100 mm by 200 mm) concrete cylinders or cores (Komar et al. 2013; Sanchez, Multon et al., 2014). The apparatus consists of a hollow cylindrical test chamber that envelops the curved surface of the test cylinder. At either end of the testing chamber, rubber "O-rings" are used to seal the compressed gas so that it only acts upon the curved surface of the specimen. Figure 6.15 illustrates the cross-section of a gas pressure chamber. Both ends are left open to atmospheric pressure, resulting in a biaxial loading configuration. Gas pressure is monotonically increased until the test cylinder fails in a plane transverse to the axis of the testing chamber (Komar et al. 2013; Sanchez, Multon et al., 2014).

The gas pressure applied to the curved surface is a biaxial loading condition, but the reaction forces within the diphase model differ. In particular, the pore water reacts hydrostatically, whereas the solid phase reacts biaxially, resulting in a net internal tensile force driven by the pore fluid. The resultant internal tension force is the primary reason why the pressure tension method is thought to measure the "pure" or "true" tensile strength of concrete, along with being well suited for detecting durability issues that affect the concrete microstructure and physical integrity (Komar et al., 2013; Sanchez, Multon et al., 2014).

Understanding the failure process in tension, along with the influence of the various tension test set-ups on the mechanical responses of "sound," but especially ISR-affected concrete, is crucial to evaluate and compare their

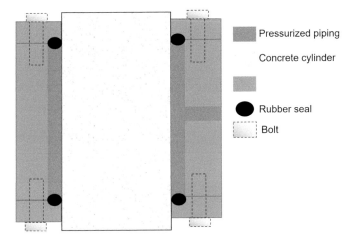

Figure 6.15 **Cross-section of the pressure chamber (Sanchez, Multon et al., 2014).**

diagnostic character. For instance, Figure 6.16 illustrates the comparison amongst the direct tensile, splitting tensile and gas pressure tension test results measured on AAR-affected concrete as a function of induced expansion. Analysing the results, it is clear to see that there is a decrease in tensile strength as a function of AAR-induced development from the beginning of the expansion development for all the set-ups presented. Moreover, the test responses vary according to the set-up used; the splitting test yields lower mechanical reductions (i.e., higher total values) as a function of induced expansion than both the direct tensile and the gas pressure tension tests, which are very close to one another. This behaviour is observed because the splitting tensile strength is an indirect test where the failure plane is previously "established" or "selected" prior to testing (i.e., vertical plane). Therefore, whether the most prominent cracks due to ISR (i.e., AAR in this case) are not located in the selected failure plane, the method is less effective in assessing the deterioration present in the specimen. Conversely, the direct tensile strength and the gas pressure tension test display a quite sharp drop in tensile strength from the beginning of the expansion development, presenting very similar results to one another as a function of the induced expansion. This indicates these two set-ups are able to measure the "pure" or "true" tensile strength of the affected concrete, creating a fracture process that is less established from the beginning of the test and thus being more effective to appraise the induced development of ISR mechanisms. Since the literature shows a number of practical issues related to the direct tensile strength test, such as failure at the ends of the specimens due to secondary stresses induced by the grips, the gas pressure tension emerges as a promising technique.

The gas pressure tension has been used to assess ISR-affected specimens and demonstrated to be suitable to capture the induced expansion and

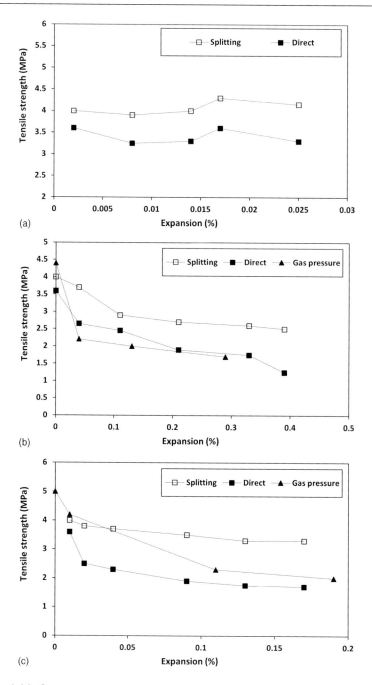

Figure 6.16 Comparison of tensile strength test results as a function of AAR-induced expansion for direct tensile and splitting tensile for (a) non-reactive aggregates, (b) highly reactive Texas sand and (c) highly reactive Québec City limestone.

(Adapted from Smaoui et al. 2006.)

damage development in its early stages (i.e., low and moderate expansion levels), regardless of the ISR mechanism (Sanchez et al., 2017, 2018). Figure 6.17 illustrates the gas pressure tension results obtained for ASR and FT-affected specimens displaying from low (i.e., 0.05%) to very high (i.e., 0.30%) expansion levels. Observing the data, one notices that the drop in

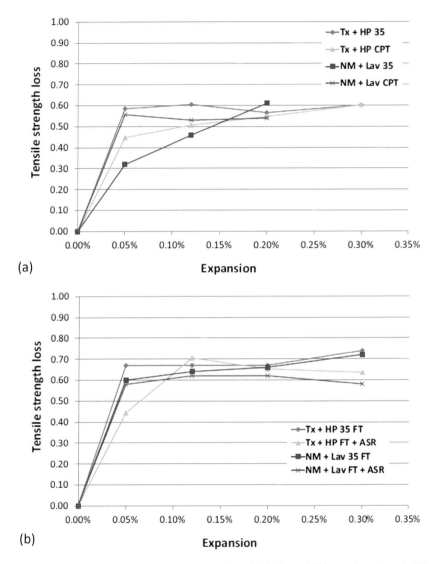

Figure 6.17 Gas pressure tension results for (a) ASR and (b) combined with FT-affected concrete. The mixtures presented in the labels are 35 MPa concrete mixtures incorporating either NM or TX as coarse or fine reactive aggregates, respectively. These mixtures developed DEF, FT and ASR deterioration.

(Adapted from Sanchez et al., 2018.)

tensile strength of FT-affected specimens (i.e., from 45% to 70%) at low expansion levels is higher than ASR-affected concrete (i.e., from 30% to 60%). Yet, for high and very high expansion levels, the difference between mechanisms seems to decrease. These results are caused by the more important presence of cracks in the cement paste (i.e., ITZ, bulk cement paste and pores) taking place in FT deterioration when compared to ASR, where the cracks are mainly within the aggregate particles in the early stages of the mechanism. However, very likely due to the test conditions (i.e., pure tension governed by fracture mechanics), the gas pressure set-up seems to not be able to distinguish different damage scenarios beyond "moderate" damage degrees since an important levelling off trend is observed from 0.12% expansion and onwards (Sanchez et al., 2017, 2018). Further analysis should still be performed to better understand the limitation of the gas pressure tension to appraise ISR-affected concrete at high and very high expansion levels since the use of this test procedure in this perspective is fairly new.

6.4 TEST PROCEDURES IN SHEAR

6.4.1 Direct shear test

Shear strength in concrete is a property governed by tension and compression forces. Once cracked (and concrete will always present a certain number of inner cracks, flaws, etc.), concrete may transfer shear forces across the cracks through two distinct mechanisms: (a) dowel effect and (b) shear friction (Haskett et al. 2011). The dowel effect is related to the reinforcement used in concrete members, whereas the shear friction is associated with the concrete features themselves. Shear friction is defined as the "frictional resistance of cracks to sliding" (Haskett et al. 2011). Under initially cracked conditions, the sliding plane surfaces of concrete may be assumed as "rough and irregular" due to the presence of aggregates (normally rough and irregular), along with the rough texture of the cracks' surfaces themselves. These rough and irregular aggregate particles "force the sliding planes apart, which induces normal stresses in the reinforcement crossing the sliding planes, restricting their opening" (Haskett et al. 2011). Confinement to the sliding planes provides frictional resistance to sliding and allows the shear forces to transfer across the cracked planes (Saouma et al. 2016; Haskett et al. 2011). Under high confinement levels, significant shear stresses are transferred across the crack surfaces through shear friction. The shear friction phenomenon, often called "aggregate interlock", is an important component while designing reinforced concrete members in North America, particularly in Canada (CSA A23.3-14 2017).

A number of test methods have been developed over the years to evaluate the direct shear capacity and shear friction of reinforced and unreinforced concrete (Banks-Sills & Arcan 1983; Adams & Walrath 1987; Richard 1981). Figure 6.18 illustrates the proposed set-ups.

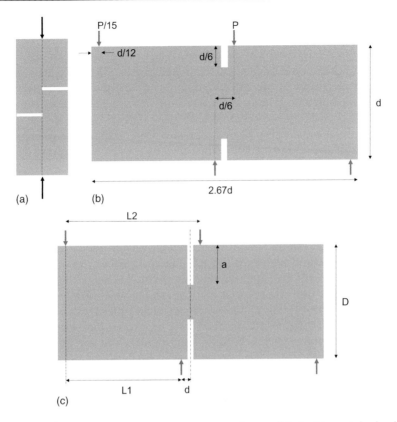

Figure 6.18 Shear set-ups to test concrete specimens: (a) double-notched cylindrical shear specimen (Barr & Hasso 1986), (b) reinforced concrete
shear set-up (Bazant & Pfeiffer 1985) and (c) small-scale direct shear
set-up (Barr & Hasso 1986).

Barr et al. (Barr & Hasso 1986) performed a comparative study on different set-ups to evaluate the direct shear of plain concrete. One of the most
promising set-ups used was the modified cylinder specimen (i.e., double-
notched cylindrical specimen), illustrated in Figure 6.18a. The proposed
geometry enabled the use and comparison of distinct types of materials and
specimens such as concrete (samples or cores made of plain concrete, fibre
reinforced concrete, etc.), rock, and similar materials (Barr & Hasso 1986).
However, some concerns were raised with the use of the double-notched
specimen, such as the difficulty of making similar and aligned notches along
with the application of a direct longitudinal load on the specimen.

Gao et al. (1979) developed a new shear set-up to study the brittle fracture
of reinforced concrete composites, which is illustrated in Figure 6.18b; later
on, this set-up was adopted for other studies with promising results. Recent
studies adapted the same geometry proposed by Gao et al., yet an upper
notch has been added to the superior part of the specimen (Gao et al. 1979).

Bazant and Pfeiffer (1985) have used the geometry proposed in Figure 6.18b to study the shear strength of concrete and mortar beams. Barr and Thomas (Barr & Derradj 1990) have used the same geometry to determine the fracture characteristics of glass-reinforced cement. However, a great deal of care was necessary during the manufacture of the test specimens to ensure a proper alignment and the same depth of the two notches. Nonetheless, the experimental programme presented interesting outcomes with low variability and high reproducibility (Barr & Derradj 1990; Barr and Hasso, 1986).

Barr and Hasso (1986) adapted the previous set-ups and proposed a test apparatus to assess small-scale concrete samples (i.e., 100 mm by 200 mm), which is illustrated in Figure 6.18c. The most important change the authors made was the use of a circumferential notch instead of two aligned ones. Studies demonstrated that circumferential notches of about 20 to 25 mm might be able to ensure the shear-type failure of the specimens without leaving a too-small area of the sample to be tested, which might cause an increase in the discrepancy of the results (Barr & Hasso 1986).

Sanchez et al. (2015a; Sanchez, Fournier, Jolin, Bedoya, et al., 2016) verified through the use of advanced microscopic analyses that AAR-induced cracks development follows a two-step process: (1) the cracks are generated within the reactive aggregate particles (fine or coarse) in the bulk concrete volume and (2) the cracks increase in length and width and eventually run out to the cement paste. The deterioration mechanism described earlier raised concerns about the shear response of concrete members affected by ISR mechanisms since shear friction (or aggregate interlock) may be drastically decreased due to AAR-induced cracks formation within the aggregate particles. Otherwise, if the aggregate interlock of concrete specimens is directly influenced by AAR, a shear test set-up might be able to capture AAR development and thus become a potentially promising tool to evaluate AAR-induced damage.

De Souza et al. (2019) evaluated a wide range of concrete mixtures deteriorated by AAR (i.e., ASR and ACR) presenting distinct mechanical properties (i.e., 25, 35 and 45 MPa) and incorporating numerous reactive aggregate types (i.e., fine and coarse) and natures (i.e., lithotypes) using the direct shear test set-up as per Barr and Hasso (Barr & Hasso 1986). Figure 6.19a illustrates the test set-up adopted, while Figure 6.19b gives a plot of the results obtained in this work. Analysing the results obtained, it is verified that there is a consistent decrease in the shear strength as a function of AAR-induced development. This decrease is not linear but very sharp at low to moderate expansion levels (i.e., 0.05% to 0.12%), where reductions from 10% to 25% are often observed, followed by a period of slight decrease at high and very high expansion levels (i.e., 0.20% to 0.30%), where reductions of 20% to 35% are verified.

Besides the promising results demonstrating the potential of the direct shear test of being a diagnostic procedure to capture AAR-induced expansion and damage development, analyses on the failure mode of tested

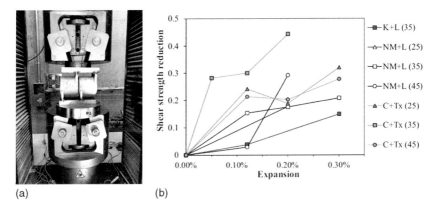

(a)   (b)

Figure 6.19 Direct shear test: (a) set-up and (b) shear strength decrease as a function of AAR-induced expansion (Souza et al., 2019).

(a)   (b)

Figure 6.20 Direct failure plane of AAR-affected concrete generated by (a) reactive fine aggregates and (b) reactive coarse aggregates. The width the image represents is 6 cm. The dashed arrows point to failure through the aggregate particles, and the solid arrows point to failure at the cement paste, causing the debonding of the aggregate particles.

(Photos courtesy of Rouzbeh Ziapour.)

specimens showed that the procedure is able to "recognize" the distinct microscopic damage features developed when AAR is triggered by reactive coarse or fine aggregates, as illustrated in Figure 6.20. For instance, the failure plane takes place through the cement paste when AAR is generated by reactive fine aggregates (Figure 6.20a) while through the coarse aggregates when AAR is triggered by reactive coarse aggregate particles (Figure 6.20b). The latter means that, first, the direct shear test set-up is able to capture AAR-induced damage, detecting distinct AAR cracking patterns and thus generating failure at distinct planes. With the promising results obtained for AAR-affected concrete, the direct shear test demonstrates a great potential to evaluate other ISR mechanisms. Further analysis should be conducted in this regard.

REFERENCES

Adams, D. F., & Walrath, D. E. (1987). Current status of the Iosipescu shear test method. *Journal of Composite Materials*, 21(6): 494–507. https://doi.org/10.1177/002199838702100601

ASTM C 469. (2006). Standard Test Method for Static Modulus of Elasticity and Poisson's Ratio of Concrete in Compression. https://doi.org/10.1520/C0469

ASTM C1293. (2015). Standard Test Method for Determination of Length Change of Concrete Due to Alkali-Silica Reaction. *ASTM Standard Book*. www.astm.org

ASTM C39. (2003). Standard test method for compressive strength of cylindrical concrete specimens. *ASTM Standard Book*, 4(March), 1–5. https://doi.org/10.1520/C0039

ASTM C42. (2008). Standard Test Method for Obtaining and Testing Drilled Cores and Sawed Beams of Concrete. *ASTM Standard Book*. https://doi.org/10.1520/mnl10913m

ASTM C496. (2008). Standard Test Method for Splitting Tensile Strength of Cylindrical Concrete Specimens. *ASTM Standard Book*. https://doi.org/10.1520/mnl10881m

ASTM C78. (2010). Standard Test Method for Flexural Strength of Concrete (Using Simple Beam with Third-Point Loading). *ASTM Standard Book*. https://doi.org/10.1520/C0078

Banks-Sills, L., & Arcan, M. (1983). An Edge-Cracked Mode II Fracture Specimen. *Experimental Mechanics*, 23(3), 257–261. https://doi.org/10.1007/bf02319251

Barr, B., & Derradj, M. (1990). Numerical study of a shear (Mode II) type test specimen geometry. *Engineering Fracture Mechanics*, 35(1–3), 171–180. https://doi.org/10.1016/0013-7944(90)90194-L

Barr, B., & Hasso, E. B. D. (1986). Development of a Compact Cylindrical Shear Test Specimen. *Journal of Materials Science Letters*, 5(12), 1305–1308. https://doi.org/10.1007/BF01729401

Bazant, Z. P., & Pfeiffer, P. A. (1985). Tests on Shear Fracture and Strain-Softening in Concrete. In *2nd Symposium on Interaction of Non-Nuclear Munitions with Structures*. Held in Pana beach, Florida, pp. 254–264.

Broek, D. (1982). *Elementary engineering fracture mechanics* (3rd ed.). Martinus Nijhoff Publishers. ISBN: 978-94-010-8425-3

Chrisp, T. M., Waldron, P., & Wood, J. G. M. (1993) Development of a non-destructive test to quantify damage in deteriorated concrete. *Magazine of Concrete Research*, 45(165), 247–256. https://doi.org/10.1680/macr.1993.45.165.247

Chrisp, T. M., Wood, J. G. M., & Norris, P. (1989). Towards quantification of microstructural damage in AAR deteriorated concrete. In *Fracture of concrete and rock: Recent developments* (p. 755).

Crouch, R. S. (1987). Specification for the determination of stiffness damage parameters from the low cyclic uniaxial compression of plain concrete cores, revision A. *Mott. Hay & Anderson. Special Services Division. Internal Technical Note.*

Crouch, R. S., & Wood, J. G. M. (1990). Damage evolution in AAR affected concretes. *Engineering Fracture Mechanics*, 35(1–3), 211–218. https://doi.org/10.1016/0013-7944(90)90199-Q

CSA A23.3-14. (2017). Design of Concrete Structures.

CSA.A23.2-14C. (2009). Obtaining and Testing Drilled Cores for Compressive Strength Testing.

Gao, H., Wang, Z., Yang, C., & Zhou, A. (1979). An investigation on the brittle fracture of KrKn composite mode cracks. *Acta Metallurgica Sinica, 15*, 380–391.

Haskett, M., Oehlers, D. J., Mohamed Ali, M. S., & Sharma, S. K. (2011). Evaluating the shear-friction resistance across sliding planes in concrete. *Engineering Structures, 33*(4), 1357–1364. https://doi.org/10.1016/j.engstruct.2011.01.013

Kongshaug, S. S., Oseland, O., Kanstad, T., Hendriks, M.A.N., Rodum, E., & Markeset, G. (2020). Experimental investigation of ASR-affected concrete – The influence of uniaxial loading on the evolution of mechanical properties, expansion and damage indices. *Construction and Building Materials*, 245, 118384. https://doi.org/10.1016/j.conbuildmat.2020.118384

Komar, A., Hartell, J., & Boyd, A. J. (2013). Pressure tension test: Reliability for assessing concrete deterioration. In *Proceedings of the Seventh International Conference on Concrete under Severe Conditions*, China, pp. 340–347.

Kosmatka, S. H., Kerkhoff, B., Panarese, W. C., MacLeod, N. F., & McGrath, R. J. 2003. *Design and control of concrete mixtures*. Canadian Cement Association (CCA).

Kubo, Y., & Nakata, M. (2012). Effect of types of reactive aggregate on mechanical properties of concrete affected by Alkali-Silica reaction. In *14th International Conference on Alkali-Aggregate Reaction*, 021711, p. 10.

Mier, J. G. M. Van, & Van Vliet, M. R. A. (2002). Uniaxial tension test for the determination of fracture parameters of concrete: State of the Art. *Engineering Fracture Mechanics, 69*, 235–247.

Mindess, S., Young, J. F., & Darwin, D. (2003). *Concrete* (2nd ed.). Prentice Hall – Person Education Inc.

Nixon, P. J., & Bollinghaus, R. 1985. The effect of Alkali aggregate reaction on the tensile strength of concrete. *Durability of Building Materials, 2*, 243–248.

Pleau, R., Bérubé, M. A., Pigeon, M., Fournier, B., & Raphael, S. (1989). Mechanical behavior of concrete Aaffected by AAR. In *8th ICAAR – International Conference on Alkali-Aggregate Reaction in Concrete* (pp. 721–726).

Richard, H. A. (1981). A new shear compact test specimen. *International Journal of Fracture, 17*, 5–7.

Sanchez, L.F.M., Drimalas, T., & Fournier, B. (2020). Assessing condition of concrete affected by Internal Swelling Reactions (ISR) through the Damage Rating Index (DRI). *Cement, 1–2*(September), 100001. https://doi.org/10.1016/j.cement.2020.100001

Sanchez, L. F. M., Drimalas, T., Fournier, B., Mitchell, D., & Bastien, J. (2018). Comprehensive damage assessment in concrete affected by different internal swelling reaction (ISR) mechanisms. *Cement and Concrete Research, 107*(February), 284–303. https://doi.org/10.1016/j.cemconres.2018.02.017

Sanchez, L.F.M., Fournier, B., Jolin, M., & Bastien, J. (2014). Evaluation of the Stiffness Damage Test (SDT) as a Tool for Assessing Damage in Concrete Due to ASR: Test Loading and Output Responses for Concretes Incorporating Fine or Coarse Reactive Aggregates. *Cement and Concrete Research*, 56. https://doi.org/10.1016/j.cemconres.2013.11.003

Sanchez, L.F.M., Fournier, B., Jolin, M., & Bastien, J. (2015b). Evaluation of the Stiffness Damage Test (SDT) as a tool for assessing damage in concrete due to Alkali-Silica Reaction (ASR): Input parameters and variability of the test responses. *Construction and Building Materials*, 77. https://doi.org/10.1016/j.conbuildmat.2014.11.071

Sanchez, L.F.M., Fournier, B., Jolin, M., Bastien, J., & Mitchell, D. 2016. Practical use of the Stiffness Damage Test (SDT) for assessing damage in concrete infrastructure affected by alkali-silica reaction. *Construction and Building Materials, 125.* https://doi.org/10.1016/j.conbuildmat.2016.08.101

Sanchez, L.F.M., Fournier, B., Jolin, M., Bedoya, M.B.A., Bastien, J., & Duchesne, J. (2016). Use of damage rating index to quantify alkali-silica reaction damage in concrete: Fine versus coarse aggregate. *ACI Materials Journal, 113(4).* https://doi.org/10.14359/51688983

Sanchez, L. F.M., Fournier, B., Jolin, M., & Duchesne, J. (2015a). Reliable quantification of AAR damage through assessment of the Damage Rating Index (DRI). *Cement and Concrete Research, 67,* 74–92. https://doi.org/10.1016/j.cemconres.2014.08.002

Sanchez, L.F.M., Fournier, B., Jolin, M., Mitchell, D., & Bastien, J. (2017). Overall Assessment of Alkali-Aggregate Reaction (AAR) in concretes presenting different strengths and incorporating a wide range of reactive aggregate types and natures. *Cement and Concrete Research, 93.* https://doi.org/10.1016/j.cemconres.2016.12.001

Sanchez, L.F.M., Multon, S., Sellier, A., Cyr, M., Fournier, B., & Jolin, M. (2014). Comparative study of a chemo-mechanical modeling for Alkali Silica Reaction (ASR) with experimental evidences. *Construction and Building Materials, 72.* https://doi.org/10.1016/j.conbuildmat.2014.09.007

Saouma, V. E.., Hariri-Ardebili, M.A., Le Pape, Y., & Balaji, R. 2016. Effect of alkali-silica reaction on the shear strength of reinforced concrete structural members. A numerical and statistical study. *Nuclear Engineering and Design, 310,* 295–310. https://doi.org/10.1016/j.nucengdes.2016.10.012

Smaoui, N., Bérubé, M. A., Fournier, B., & Bissonnette, B. (2004b). Influence of specimen geometry, orientation of casting plane, and mode of concrete consolidation on expansion due to ASR. *Cement, Concrete and Aggregates, 26(2),* 58–70. https://doi.org/10.1520/cca11927

Smaoui, N., Bissonnette, B., Bérubé, M.A., Fournier, B., & Durand, B. (2006). Mechanical properties of ASR-affected concrete containing fine or coarse reactive aggregates. *Journal of ASTM International, 3(3),* 1–16. https://doi.org/10.1520/jai12010

Smaoui, N., Fournier, B., Bérubé, M. A., Bissonnette, B., & Durand, B. 2004a. Evaluation of the expansion attained to date by concrete affected by alkali-silica reaction. Part II: Application to nonreinforced concrete specimens exposed outside. *Canadian Journal of Civil Engineering, 31(6):* 997–1011. https://doi.org/10.1139/L04-074

Souza, D. J. De, Sanchez, L. F.M., & De Grazia, M. T. (2019). Evaluation of a direct shear test setup to quantify AAR-induced expansion and damage in concrete. *Construction and Building Materials, 229,* 116806. https://doi.org/10.1016/j.conbuildmat.2019.116806

Swamy, R.N. (1979). Fracture mechanisms applied to concrete. In F.D. Lydon (Eds.), *Fracture measurements of cementitious composites* (pp. 221–281). London: Applied Science Publishers.

Walsh, J. B. (1965). The effect of cracks on the uniaxial elastic compression of rocks. *Journal of Geophysical Research, 70(2),* 399–411. https://doi.org/10.1029/jz070i002p00399

Zheng, W., Kwan, A. K.H., & Lee, P.K.K. (2001). Direct tension test of concrete. *ACI Materials Journal, 98* (1): 63–71. https://doi.org/10.14359/10162

Chapter 7

Multi-level assessment of ISR-affected concrete

7.1 INTRODUCTION

In the preceding chapters, an extensive discussion unfolded on the distinct internal swelling reaction (ISR) mechanisms inducing deterioration in concrete, the importance of assessing the condition of ISR-affected concrete and the different test procedures (i.e., visual, non-destructive, chemical, microscopic and mechanical) enabling the appraisal of structures and structural members affected by ISR. The analysis of these chapters revealed the presence of advantages and disadvantages associated with each recommended technique, necessitating a case-by-case evaluation. Depending on factors such as the type of structure, accessibility, extent of damage and environmental conditions, certain techniques demonstrate greater efficiency than others. However, it became evident from the preceding chapters that the most effective approach lies in combining these techniques to create comprehensive evaluation protocols. The selection of techniques should go beyond mere "data gathering and correlation" and instead focus on their synergistic nature, providing supplementary information necessary for a thorough diagnosis (i.e., cause(s) and extent) of the damage affecting concrete structures or structural members under analysis. Chapter 5 extensively discussed the fact that different deterioration mechanisms associated with ISR can induce distinct microscopic damage features, such as cracks, in affected concrete; Figure 7.1 illustrates the various damage features associated with different ISR mechanisms.

Analysis of Figure 7.1 reveals that ASR-induced deterioration predominantly originates within the reactive aggregate particles (i.e., either fine or coarse) in the concrete microstructure. In contrast, DEF-induced deterioration often manifests as cracks outlining the aggregate particles at the interfacial transition zone (ITZ) and subsequently extending into the cement paste, ultimately connecting with one another. In the case of freeze and thaw (FT) deterioration, cracks tend to develop primarily in the cement paste and macro pores (assuming the use of FT-resistant aggregates in the mixture) rather than exclusively at the ITZ. Moreover, it is worth noting that the proximity of FT cracks to the surface of affected concrete members

DOI: 10.1201/9781003188155-7

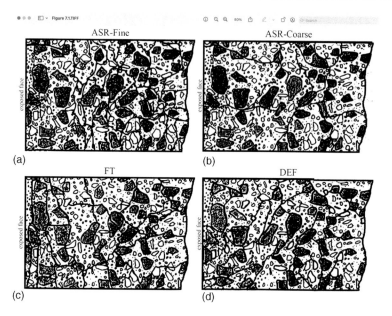

Figure 7.1 Crack pattern in concrete affected by (a) alkali-silica reaction (ASR) reactive coarse aggregate, (b) ASR reactive sand, (c) freezing and thawing cycles and (d) delayed ettringite formation (DEF). The concrete's surface is shown on the left-hand side.

(Adapted from British Cement Association (BCA) 1992.)

correlates with a higher likelihood of parallel alignment. These observations emphasize the fact that the overall impact of deterioration on the engineering properties of concrete varies depending on the underlying mechanism(s) causing damage and, significantly, the degree of its development. Therefore, an efficient and comprehensive assessment protocol should encompass the ability to determine these variations.

7.2 MULTI-LEVEL ASSESSMENT PROTOCOL

Literature reports a wide range of test procedures (e.g., visual, non-destructive, microscopic and mechanical) employed to quantify the impact of ISR-induced deterioration in affected concrete over time. However, most of the available data is presented through a "decoupled" pathway. This means that either microscopic and descriptive data are provided to confirm the presence of ISR through the identification of secondary products or visual, non-destructive and mechanical data shed light on the influence of current ISR-induced deterioration on the durability and engineering properties of the affected material. The limitation of conducting "decoupled" analyses is the inability to determine the cause and extent (i.e., complete diagnosis) of the deterioration. Furthermore, without establishing a quantitative correlation between the "inner quality" of the affected concrete microstructure and its macro performance, the potential progression of the induced damage and its associated impact remains elusive. To overcome this challenge, *a multi-level assessment protocol*, recognized as a multi-scale approach, has been developed to facilitate a more comprehensive understanding, correlation and quantitative description of how the "inner quality" of the concrete microstructure (i.e., porosity, number, type and location of cracks; presence of secondary products; discoloration; and other relevant factors) influences its macro performance (Martin et al. 2017; Sanchez et al. 2017, 2018). Essentially, the macro performance of concrete, comprehending engineering properties, such as stiffness, compressive, tensile and direct shear strength, is explained over time via a thorough and quantitative examination of its microstructure. The multi-level protocol recommends the integration of quantitative microscopic techniques, such as the Damage Rating Index (DRI) and image analysis, combined with mechanical tests, such as the stiffness damage test (SDT) and direct shear test, to establish a cohesive relationship between the microstructure and macro performance. Over the years, this protocol has exhibited a high efficiency and reliability in assessing the condition of concrete damaged by ISR, thereby enabling a deeper comprehension of its present state and the progressive advancement of induced deterioration in ISR-affected critical concrete infrastructure.

7.3 PRACTICAL USE OF THE MULTI-LEVEL ASSESSMENT PROTOCOL

In the pursuit of a comprehensive understanding of ISR-affected concrete, Sanchez et al. and Martin et al. (Martin et al. 2017; Sanchez et al. 2017, 2018) conducted an extensive experimental campaign using numerous concrete specimens/cores with distinct raw materials and compositions. As a result, a pragmatic multi-level approach was proposed, which integrates quantitative microscopic and mechanical data into a unified four-quadrant chart (Figure 7.2).

Within this chart, the expansion experienced by ISR-affected concrete at a specific age (represented on the positive x-axis) can be easily correlated with mechanical data (expressed as Stiffness Damage Index (SDI) on the positive y-axis) and microscopic data (DRI number divided by 1,000 on the negative y-axis). Moreover, additional parameters of interest, such as modulus of elasticity (ME), compressive strength, direct shear strength and more, can be incorporated on the negative x-axis, thereby enabling a comprehensive evaluation of the overall degree of deterioration.

The compilation and analysis of data through the four-quadrant chart for ISR-affected concrete, incorporating a wide variety of raw materials and compositions, enabled the development of benchmark tables that outline

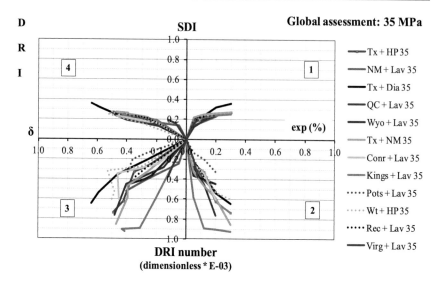

Figure 7.2 Four-quadrant chart for the assessment of ISR-affected concrete (Sanchez et al. 2017, 2018).

the impact of different ISR mechanisms on the induced expansion and deterioration of affected concrete. Table 7.1 presents examples of such benchmark tables for ASR, DEF and FT mechanisms. Notably, this experimental campaign also investigated induced deterioration caused by alkali-carbonate reaction (ACR) (i.e., Kings + Lav 35), which allowed a deeper understanding of its distinct impact on concrete integrity and engineering properties. However, it is important to note that ACR was not included in the benchmark tables due to the utilization of only one aggregate source for this particular mechanism.

The utilization of benchmark tables holds considerable advantages for engineers in practice, as it enhances decision-making by providing a more accurate understanding of the entire deterioration process with a reduced reliance on extensive data collection. However, it is important to note that these benchmark tables have primarily been established based on laboratory test samples subjected to free-induced expansion conditions, while it is evident that structures and structural members experience a diverse range of stress-state conditions in the field. Nevertheless, this approach should be regarded as a "reference" and can be considered the "worst-case scenario" for ISR-affected concrete.

7.3.1 Using the multi-level assessment protocol to understand the impact of ISR under unrestrained conditions

The evaluation of ISR-affected concrete commonly relies on quantifying the degree of expansion, which serves as a measure of its deterioration.

Table 7.1 Benchmark tables of ISR-induced development on affected concrete

Distress mechanism	Classification of ASR damage degree (%)	Reference expansion level (%)[a]	Damage results				
			Stiffness loss (%)	Compressive strength loss (%)	Tensile strength loss (%)	SDI	DRI
ASR	Negligible	0.00–0.03	–	–	–	0.06–0.16	100–155
	Marginal	0.04 ± 0.01	5–37	(–)10–15	15–60	0.11–0.25	210–400
	Moderate	0.11 ± 0.01	20–50	0–20	40–65	0.15–0.31	310–500
	High	0.20 ± 0.01	35–60	13–25	45–80	0.19–0.32	500–765
	Very high	0.30 to 0.50 ± 0.01	40–67	20–35	–	0.22–0.36	600–925
	Ultra high	0.50 to 1.00 ± 0.01	–	–	–	–	–
		≥ 1.00 ± 0.01	–	–	–	–	–
FT and FT + ASR	Negligible	0.00–0.03	–	–	–	0.11	147–154
	Marginal	0.04 ± 0.01	0.23–0.35	0.12–0.32	0.44–0.67	0.16–0.23	496–684
	Moderate	0.11 ± 0.01	0.28–0.36	0.21–0.32	0.62–0.67	0.25–0.28	590–950
	High	0.20 ± 0.01	0.33–0.46	0.22–0.37	0.65–0.67	0.27–0.41	677–963
	Very high	0.30 to 0.50 ± 0.01	0.37–0.52	0.24–0.40	0.65–0.73	0.34–0.45	800–1300
	Ultra high	0.50 to 1.00 ± 0.01	–	–	–	–	–
		≥ 1.00 ± 0.01	–	–	–	–	–
DEF and DEF + ASR	Negligible	0.00–0.03	–	–	–	0.11	110–147
	Marginal	0.04 ± 0.01	–	–	–	–	–
	Moderate	0.11 ± 0.01	0.35–0.56	0.09–0.34	–	0.17–0.20	355–599
	High	0.20 ± 0.01	–	–	–	–	–
	Very high	0.30 to 0.50 ± 0.01	0.55–0.62	0.29–0.43	–	0.19–0.28	623–710
	Ultra high	0.50 to 1.00 ± 0.01	0.56–0.77	0.40–0.47	–	0.27–0.43	828–1022
		≥ 1.00 ± 0.01	0.60–0.86	0.40–0.50	–	0.30–0.54	841–1363

[a] These levels of expansion should not be considered as strict limits between the various of damage degree but rather indicators/reference levels of the three different mechanisms studied in this work.

Source: Adapted from Sanchez et al. (2017, 2018)

Extensive investigations have revealed a consistent and progressive correlation between expansion and the overall deterioration of the concrete (Mohammadi et al. 2020; Sanchez et al. 2017). Specifically, as the expansion increases, there is a corresponding escalation in the loss of mechanical properties, physical integrity and stiffness of the affected material. This correlation emphasizes the importance of accurately assessing and quantifying the extent of expansion to effectively gauge the severity of deterioration in ISR-affected concrete. In this context, extensive investigations have been conducted to understand the particularities associated with different ISR mechanisms, including ASR, ACR, DEF and FT cycles.

7.3.1.1 Alkali-silica reaction (ASR)

In an extensive investigation conducted by (Sanchez et al. 2017), the deterioration caused by ASR was thoroughly examined. The study involved the assessment of 20 different concrete mixtures, each characterized by distinct mechanical properties (e.g., 25, 35 and 45 MPa) and incorporating a total of 13 different lithotypes of reactive aggregates. The integrity of the concrete was evaluated using the DRI, while the mechanical properties were assessed through the SDT, along with measurements of ME, tensile and compressive strength.

Remarkably, the investigation revealed a pronounced correlation between the development of distress features and changes in the mechanical properties of the concrete specimens in relation to the extent of ASR-induced expansion. These significant findings are depicted in Figure 7.3.

For negligible (0.00%) and marginal levels (0.05%) of expansion, a pronounced decrease in ME (up to 30%) and tensile strength (30% to 70%) is clearly observed. It is worth noting that the majority of ASR-induced cracks are localized within the aggregate particles at these expansions (Goltermann 1995; Sanchez et al. 2017), as extensively discussed in Chapters 2 and 5. As elaborated in Chapter 6, this significant decline in concrete's ME can be attributed to the reduced stiffness of the aggregates caused by the generation of cracks at the submicroscopic level within the aggregate particles. Similarly, the notable reduction in tensile strength can be attributed to the principles of fracture mechanics (Sanchez et al. 2017; Zahedi et al. 2021). Moreover, Chapter 6 thoroughly examines the failure mechanism of concrete under compression, which elucidates the fact that ASR-induced deterioration only slightly impacts the compressive strength of affected concrete (i.e., about 5% on average) at this initial stage of expansion.

As ASR progresses and attains a moderate level of expansion (0.12%), new cracks begin to form within the aggregate particles, while pre-existing ones extend in length and width, gradually infiltrating the surrounding cement paste (Sanchez et al. 2017; Zahedi et al. 2021). Consequently, the decline in both the ME and tensile strength of the affected concrete continues, albeit at a more gradual pace (up to 35% and 55%–75%, respectively).

Notably, the compressive strength experiences a minor decrease of approximately 10% at this stage.

At higher levels of expansion (0.20%), the ASR-induced deterioration primarily propagates existing cracks within the cement paste rather than generating new ones. Chapter 6 provides detailed insights into this phenomenon. Although the ME of the concrete continues to decline at a slower rate, the

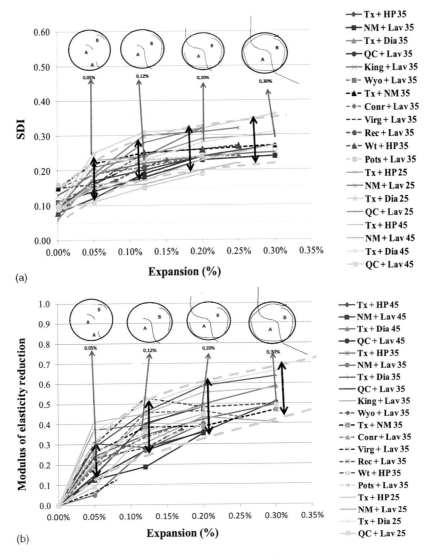

Figure 7.3 Losses in mechanical properties of ASR-affected concrete incorporating distinct aggregate types versus induced expansion: (a) SDI, (b) modulus of elasticity reduction.

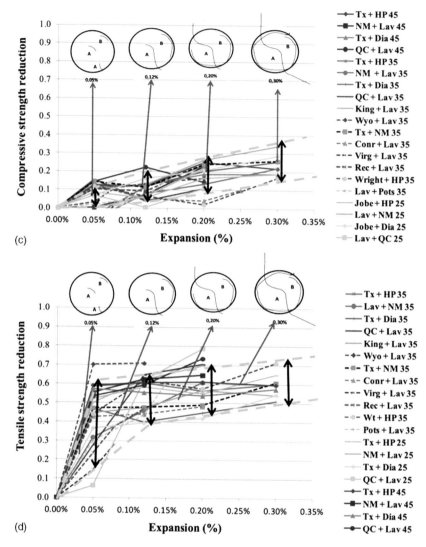

Figure 7.3 (Continued) Losses in mechanical properties of ASR-affected concrete incorporating distinct aggregate types versus induced expansion: (c) compressive strength reduction and (d) tensile strength reduction.

(Adapted from Sanchez et al. 2017, 2018.)

loss in tensile strength appears to be similar to that observed at the moderate expansion levels (0.12%). In contrast, the compressive strength of the affected concrete begins to decline more significantly, experiencing a reduction of around 25%. Ultimately, when ASR reaches very high levels of expansion (≥ 0.30%), a complex network of cracks permeates the entire cement paste, leading to a substantial drop in the compressive strength of the affected concrete (35%–40%). However, the loss in both ME and tensile strength appears to level off (Sanchez et al. 2017).

7.3.1.2 Alkali-carbonate reaction (ACR)

In the same experimental investigation conducted as part of the ASR section, an assessment of ACR-induced development was performed on 35 MPa concrete mixtures incorporating a highly reactive coarse aggregate sourced from Kingston, Ontario, Canada (Sanchez et al. 2017). ACR-induced development was monitored over time, and the concrete microstructure's analysis through the DRI, along with the evaluation of ACR's impact on the mechanical properties of affected concrete via SDT, ME, tensile and compressive strength, were conducted. While no descriptive deterioration model has been proposed to fully elucidate the development of ACR-induced deterioration, it is evident that the microscopic damage features resulting from ACR-induced expansion differ significantly from those observed in the case of ASR crack development (see Figure 7.4).

During the initial stages of the chemical reaction and at lower levels of expansion (0.05%), the observations revealed the presence of a few closed and open cracks within the aggregate particles. However, it is noteworthy that significant cracking without secondary products (i.e., gel) formation was already detected in the cement paste, primarily concentrated at the ITZ. The mechanism behind this type of cracking at the ITZ can be attributed to two distinct phenomena – namely, expansion of the cement paste (which is unlikely in this case) or shrinkage of the aggregate particles, as extensively discussed in Chapter 2.

For moderate levels of expansion (0.12%), the network of cracking within the cement paste continues to intensify, ultimately resulting in a substantial density of cracks in the affected concrete. Although a very different cracking pattern is observed when compared to ASR-induced deterioration, the reduction in mechanical properties seems to be quite similar; however, the reduction in compressive strength of ACR-affected concrete is more pronounced since the primary cracks generated in the mechanism are located within the ITZ and bulk cement paste.

7.3.1.3 Delayed ettringite formation (DEF)

Experimental campaigns have been conducted to investigate the implications of DEF on the integrity of concrete and its subsequent reduction in mechanical properties (E. Giannini et al. 2018; Melo et al. 2023; Sanchez et al. 2018). The findings obtained from these experimental programmes shed some light on the crack generation and progression within the ITZ during the initial stages of expansion (up to 0.12%). Notably, the presence of cracks in the ITZ adversely affects the bond between the aggregate and

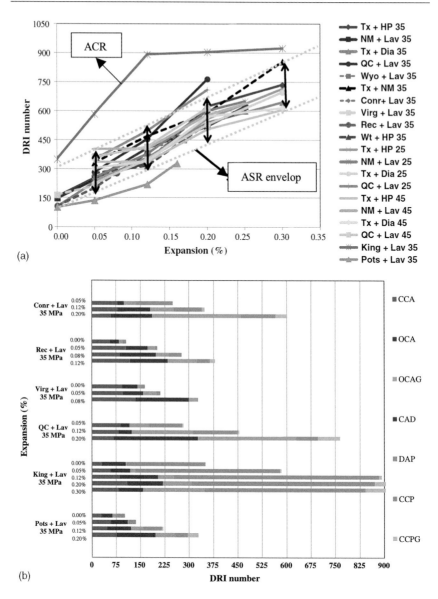

Figure 7.4 Comparison of DRI between ACR and ASR: (a) plot and (b) bar charts (Sanchez et al. 2017).

cement paste. Consequently, a significant decrease in the ME, reaching up to 50%, is observed. Moreover, the reduction in compressive strength, averaging around 10%, is relatively higher when compared to ASR-affected concrete. The relatively limited decrease in compressive strength at this stage of the reaction, despite the majority of cracks being present in the ITZ, can be attributed to either the highly localized nature of DEF-induced cracking or the potential inhibitory effect caused by the "stop/arrest mechanism" provided by aggregate particles, as previously proposed by (Mehta & Monteiro 2014; Sanchez et al. 2018).

As the expansion level increases to a higher range (0.30–0.50%), in accordance with the minimum energy law, the pre-existing cracks in the system become more significant in terms of length and width, while the formation of new cracks becomes limited. Cracks within the ITZ start to interconnect, giving rise to a network of cracks within the cement paste. Consequently, the ME continues to decrease, reaching a reduction of up to 60%, while the compressive strength is notably diminished by up to 40%. At very high levels of expansion (> 0.50%), the aggregate particles undergo debonding and disaggregation, leading to a substantial decrease in the ME of the affected concrete, with reductions of up to 85%. Likewise, the compressive strength experiences a significant decline of up to 50% due to the formation of a cracking network within the bulk cement paste (Sanchez et al. 2018) (Figure 7.5).

7.3.1.4 Freeze and thawing (FT)

Research has been conducted to understand FT-induced expansion and deterioration and its impact on the mechanical properties of affected concrete. This understanding is crucial to understand the behaviour of concrete under harsh climates (Sanchez et al. 2018, 2020; Zahedi et al. 2022a). The findings obtained allowed the description of FT-induced deterioration (i.e., generation and propagation) as a function of expansion, as follows.

At the initial stage of damage (0.05%), FT-induced cracks are observed in either the ITZ or the bulk cement paste. While the ME does not appear to be significantly affected when compared to ASR, the tensile strength of the concrete at this stage matches that of ASR-affected concrete, with a reduction of approximately 60% (Komar & Boyd 2017). However, it is worth noting that FT-induced cracks are narrower and more widespread when compared to DEF-induced cracks. Consequently, a slightly higher loss of compressive strength is observed at this stage for FT deterioration, ranging between 15% and 30% (Sanchez et al. 2018).

As the expansion progresses to 0.12%, there is no significant change in the ME and tensile strength when compared to the previous stage of damage. However, the compressive strength continues to decrease, reaching a reduction of approximately 20% to 35%. This decrease can be attributed to the propagation of most FT-induced cracks into the cement paste, either within the ITZ or the bulk paste/pores.

At higher levels of expansion (0.20%), the FT cracks continue to propagate, resulting in the interconnection of various cracks in the cement paste, as well as some cracks within the aggregate particles. This leads to the formation of an important cracking network, which significantly diminishes the physical integrity of the affected material.

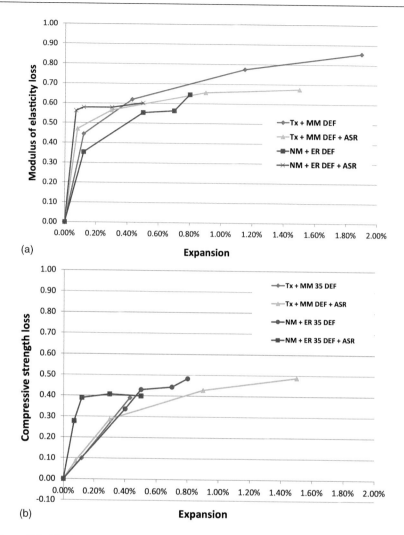

Figure 7.5 Mechanical properties losses DEF-affected concrete specimens incorporating distinct aggregate types as a function of concrete expansions: (a) modulus of elasticity and (b) compressive strength.

Finally, at very high expansion levels (0.30%), the loss of ME (approximately 50%) becomes almost identical to that of ASR-affected specimens, although still lower than DEF-induced deterioration (Sanchez et al. 2018). This behaviour may be attributed to the observation of the aggregate particles' debonding in DEF deterioration, as reported by (BCA 1992; Poole & Sims 2016; Thomas et al. 2008). Furthermore, at this expansion level, a drop in compressive strength of around 40% to 45% is experienced, which is nearly identical to DEF and slightly higher than ASR-affected concrete under similar levels of damage (Sanchez et al. 2018) (Figure 7.6).

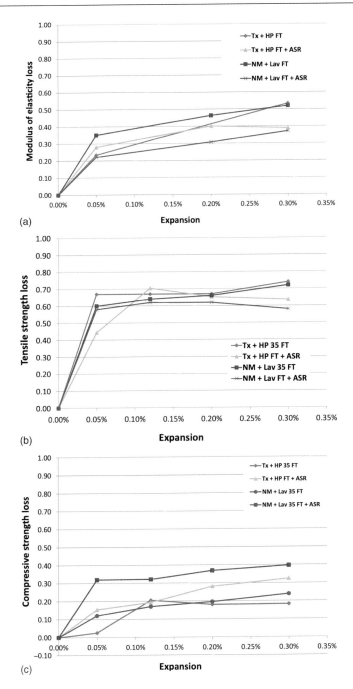

Figure 7.6 Mechanical properties losses FT-affected concrete specimens incorporating distinct aggregate types as a function of concrete expansions: (a) modulus of elasticity, (b) tensile strength and (c) compressive strength.

7.3.2 Using the multi-level assessment protocol to understand the impact of ISR under restrained conditions

Extensive research has been conducted to evaluate the behaviour of various ISR mechanisms under unrestrained (free expansion) conditions. However, only a limited number of studies have investigated the influence of confinement on ISR-induced expansion and deterioration (Zahedi et al., 2022c, 2022d). The majority of these studies have concurred that restraint helps mitigate ISR-induced expansion (Multon & Toutlemonde 2006). It has also been observed the anisotropic character of ISR under restraint since its induced expansion and deterioration are diminished in the direction of confinement while being transferred (and thus increasing) to unrestrained directions (Morenon et al. 2017; Multon et al. 2005; Multon & Toutlemonde 2006; Zahedi et al., 2022c, 2022d). For instance, previous observations have verified that unrestrained directions of reinforced concrete blocks reinforced in a single dimension (i.e., 1D) with a reinforcement ratio of 2% exhibited approximately 40% higher ISR-induced expansion when compared to the confined direction (Zahedi et al. 2021; Zahedi et al. 2022b, 2022c, 2022d). This suppression of induced expansion in the main reinforcement direction is attributed to the elastic restraint of the reinforcement, which induces tensile stresses in the steel rebars.

On the one hand, anisotropy plays an important role in the deformation of ISR-affected members; an interesting example was observed in the study of thick slabs bearing rectangular cross-sections and asymmetrically distributed longitudinal reinforcement around their centroidal axis; it has been verified that the original rectangular cross-sections deformed into trapezoidal sections after approximately 0.15%–0.20% of induced expansion (Allard et al. 2018). Schematics illustrating such deformations can be found in Figure 7.7. On the other hand, anisotropy can significantly influence the orientation of induced cracks based on the boundary conditions and reinforcement configuration of affected concrete members. While unconfined members typically uncover a random crack pattern (i.e., map cracking), most ISR-induced cracks in restrained concrete tend to be generated parallel to the main reinforcement (Allard et al. 2018; Barbosa et al. 2018; Zahedi et al., 2022c, 2022d). Schematics depicting the development of ISR-induced cracks in unrestrained, 1D and 2D reinforced concrete blocks can be found in Figure 7.8 (Zahedi et al. 2021).

In addition to influencing crack orientation, anisotropy can also impact the generation and propagation of cracks in reinforced concrete. Compared to concrete under free expansion conditions, ISR-affected reinforced concrete exhibits a higher number of open cracks in the cement paste, along with evidencing a higher amount of secondary products/deposits for the same level of development, particularly in ASR deterioration. Moreover, the

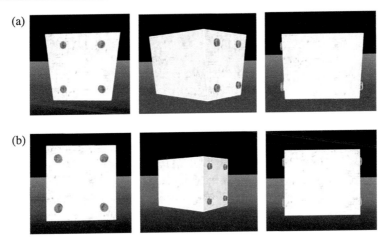

Figure 7.7 Schematics depicting (a) damaged ASR-affected restrained concrete block and (b) undamaged concrete block (Zahedi et al. 2022b).

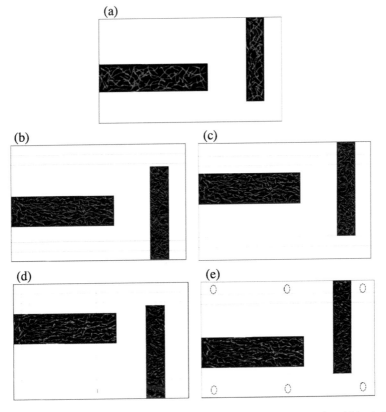

Figure 7.8 Typical ASR-induced crack orientation from (a) unconfined block (top view), (b) 1D confined block (top view), (c) 1D confined block (side/longitudinal view), (d) 2D confined block (top view) and (e) 2D confined block (side/longitudinal view) (Zahedi et al. 2021).

generation of an important number of cracks in non-reactive aggregate particles is often observed (Zahedi et al. 2022b).

The observation of such singular petrographic damage features in reinforced concrete can lead to the identification of different mechanical property losses in restrained vs unrestrained members. The degradation of the ME and compressive strength in ISR-affected reinforced concrete follows similar anisotropic trends (Abd-Elssamd et al. 2020; Giannini & Folliard 2012; Hayes et al. 2020), and higher reductions in compressive strength and ME are observed in unrestrained directions when compared to the restrained directions (Barbosa et al. 2018; Zahedi et al., 2022c, 2022d). For instance, the compressive strength of cores extracted perpendicular to the main restrained direction may experience higher reductions ranging from 7% to 35% when compared to those retrieved parallel to the main rebars (Bach et al. 1993; Barbosa et al. 2018; Jones et al. 1994; Kongshaug et al. 2020; Morenon et al. 2017). This behaviour matches the number of cracks observed in the cores, where cores parallel to the steel bars tend to display a slightly lower number of cracks when compared to those extracted perpendicular to the rebars (Zahedi et al., 2022c, 2022d).

Ultimately, when comparing the impact of various confinement configurations (i.e., none, uniaxial, biaxial and triaxial) on ISR-induced damage development, it is observed that concrete members under triaxial conditions exhibit lower expansion, microscopic damage and mechanical degradation than the other restraint configurations over time.

REFERENCES

Abd-Elssamd, A., Ma, Z. J., Le Pape, Y., Hayes, N. W., & Guimaraes, M. (2020). Effect of alkali-silica reaction expansion rate and confinement on concrete degradation. *ACI Materials Journal*, *117*(1). https://doi.org/10.14359/51720294

Allard, A., Bilodeau, S., Pissot, F., Fournier, B., & Bissonnette, B. (2018). Expansive behavior of thick concrete slabs affected by alkali-silica reaction (ASR). *Construction and Building Materials*, *171*, 421. https://doi.org/10.1016/j.conbuildmat.2018.03.111

Bach, F., Thorsen, T. S., & Nielsen, M. P. (1993). Load-carrying capacity of structural members subjected to alkali-silica reactions. *Construction and Building Materials*, *7*(2), 109–115. https://doi.org/10.1016/0950-0618(93)90040-J

Barbosa, R. A., Hansen, S. G., Hansen, K. K., Hoang, L. C., & Grelk, B. (2018). Influence of alkali-silica reaction and crack orientation on the uniaxial compressive strength of concrete cores from slab bridges. *Construction and Building Materials*, *176*, 440–451. https://doi.org/10.1016/j.conbuildmat.2018.03.096

British Cement Association (BCA). (1992). *The diagnosis of alkali-silica reaction* (p. 44) [Report of a Working Party]. British Cement Association (BCA).

Giannini, E., & Folliard, K. (2012). Stiffness damage and mechanical testing of core specimens for the evaluation of structures affected by ASR. *14th International Conference on Alkali-Aggregate Reaction*. International Conference on Alkali-Aggregate Reaction, Texas, USA.

Giannini, E., Sanchez, L. F. M., Tuinukuafe, A., & Folliard, K. J. (2018). Characterization of concrete affected by delayed ettringite formation using the stiffness damage test. *Construction and Building Materials*, 162, 253–264. https://doi.org/10.1016/j.conbuildmat.2017.12.012

Goltermann, P. (1995). Mechanical predictions of concrete deterioration; Part 2: Classification of crack patterns. *ACI Materials Journal*, 92(1). https://doi.org/10.14359/1177

Hayes, N. W., Giorla, A. B., Trent, W., Cong, D., Le Pape, Y., & Ma, Z. J. (2020). Effect of alkali-silica reaction on the fracture properties of confined concrete. *Construction and Building Materials*, 238, 117641. https://doi.org/10.1016/j.conbuildmat.2019.117641

Jones, A. E. K., Clark, L. A., & Amasaki, S. (1994). The suitability of cores in predicting the behaviour of structural members suffering from ASR. *Magazine of Concrete Research*, 46(167), 145–150. https://doi.org/10.1680/macr.1994.46.167.145

Komar, A. J. K., & Boyd, A. J. (2017). Evaluating freeze-thaw deterioration with tensile strength. *IOP Conference Series: Materials Science and Engineering*, 216, 012024. https://doi.org/10.1088/1757-899X/216/1/012024

Kongshaug, S. S., Oseland, O., Kanstad, T., Hendriks, M. A. N., Rodum, E., & Markeset, G. (2020). Experimental investigation of ASR-affected concrete – The influence of uniaxial loading on the evolution of mechanical properties, expansion and damage indices. *Construction and Building Materials*, 245, 118384. https://doi.org/10.1016/j.conbuildmat.2020.118384

Martin, R.-P., Sanchez, L., Fournier, B., & Toutlemonde, F. (2017). Evaluation of different techniques for the diagnosis & prognosis of Internal Swelling Reaction (ISR) mechanisms in concrete. *Construction and Building Materials*, 156, 956–964. https://doi.org/10.1016/j.conbuildmat.2017.09.047

Mehta, P. K., & Monteiro, P. J. M. (2014). *Concrete: Microstructure, properties, and materials* (4th ed.). New York: McGraw-Hill Education.

Melo, R. H. R. Q., Hasparyk, N. P., & Tiecher, F. (2023). Assessment of concrete impairments over time triggered by DEF. *Journal of Materials in Civil Engineering*, 35(8). https://doi.org/10.1061/JMCEE7.MTENG-15041

Mohammadi, A., Ghiasvand, E., & Nili, M. (2020). Relation between mechanical properties of concrete and alkali-silica reaction (ASR); a review. *Construction and Building Materials*, 258, 119567. https://doi.org/10.1016/j.conbuildmat.2020.119567

Morenon, P., Multon, S., Sellier, A., Grimal, E., Hamon, F., & Bourdarot, E. (2017). Impact of stresses and restraints on ASR expansion. *Construction and Building Materials*, 140, 58–74

Multon, S., Seignol, J.-F., & Toutlemonde, F. (2005). Structural behavior of concrete beams affected by alkali-silica reaction. *ACI Materials Journal*, 102(2), 67–76.

Multon, S., & Toutlemonde, F. (2006). Effect of applied stresses on alkali-silica reaction-induced expansions. *Cement and Concrete Research*, 36(5), 912–920. https://doi.org/10.1016/j.cemconres.2005.11.012

Poole, A., & Sims, I. (2016). *Concrete petrography: A handbook of investigative techniques* (2nd ed.). CRC Press, Taylor & Francis Group, a Balkema book; Gale Academic OneFile. https://doi.org/10.1201/b18688

Sanchez, L. F. M., Drimalas, T., & Fournier, B. (2020). Assessing condition of concrete affected by internal swelling reactions (ISR) through the Damage Rating Index (DRI). *Cement*, 1–2, 100001. https://doi.org/10.1016/j.cement.2020.100001

Sanchez, L. F. M., Drimalas, T., Fournier, B., Mitchell, D., & Bastien, J. (2018). Comprehensive damage assessment in concrete affected by different internal swelling reaction (ISR) mechanisms. *Cement and Concrete Research*, *107*, 284–303. https://doi.org/10.1016/j.cemconres.2018.02.017

Sanchez, L. F. M., Fournier, B., Jolin, M., Mitchell, D., & Bastien, J. (2017). Overall assessment of Alkali-Aggregate Reaction (AAR) in concretes presenting different strengths and incorporating a wide range of reactive aggregate types and natures. *Cement and Concrete Research*, *93*, 17–31. http://dx.doi.org/10.1016/j.cemconres.2016.12.001

Thomas, M., Folliard, K., Drimalas, T., & Ramlochan, T. (2008). Diagnosing delayed ettringite formation in concrete structures. *Cement and Concrete Research*, *38*(6), 841–847. https://doi.org/10.1016/j.cemconres.2008.01.003

Zahedi, A., Komar, A., Sanchez, L. F. M., & Boyd, A. J. (2022a). Global assessment of concrete specimens subjected to freeze-thaw damage. *Cement and Concrete Composites*, *133*, 104716. https://doi.org/10.1016/j.cemconcomp.2022.104716

Zahedi, A., Sanchez, L. F. M., & Noël, M. (2022b). Appraisal of visual inspection techniques to understand and describe ASR-induced development under distinct confinement conditions. *Construction and Building Materials*, *323*, 126549. https://doi.org/10.1016/j.conbuildmat.2022.126549

Zahedi, A., Trottier, C., Sanchez, L. F. M., & Noël, M. (2021). Microscopic assessment of ASR-affected concrete under confinement conditions. *Cement and Concrete Research*, *145*, 106456. https://doi.org/10.1016/j.cemconres.2021.106456

Zahedi, A., Trottier, C., Sanchez, L. F. M., & Noël, M. (2022c). Condition assessment of alkali-silica reaction affected concrete under various confinement conditions incorporating fine and coarse reactive aggregates. *Cement and Concrete Research*, *153*, 106694. https://doi.org/10.1016/j.cemconres.2021.106694

Zahedi, A., Trottier, C., Sanchez, L., & Noël, M. (2022d). Evaluation of the induced mechanical deterioration of alkali-silica reaction affected concrete under distinct confinement conditions through the stiffness damage test. *Cement and Concrete Composites*, *126*, 104343. https://doi.org/10.1016/j.cemconcomp.2021.104343

Chapter 8

Forecasting future performance and managing critical infrastructure

8.1 INTRODUCTION

The management of internal swelling reaction (ISR)–affected concrete infrastructure comprises three main aspects: (a) diagnosis, (b) prognosis and (c) maintenance strategies. The diagnosis task aims to find the main cause(s) leading to deterioration and appraises the current condition (i.e., damage degree) of affected structures or structural members under evaluation. Otherwise, the prognosis consists of assessing the potential for further deterioration and the associated structural implications over time. Finally, the outcomes gathered from diagnosis and prognosis are critical for determining appropriate maintenance strategies.

Chapters 1 and 2 presented in detail the most common ISR mechanisms inducing expansion and deterioration in concrete and the need for a proper diagnosis, whereas Chapters 3 to 7 displayed numerous diagnosis techniques and approaches (i.e., visual, non-destructive, microscopic and mechanical tests), enabling a reliable appraisal of the main cause(s) and extent of deterioration of affected concrete. This chapter aims to discuss the most common prognosis procedures, frameworks and models adopted in the literature to evaluate the potential of further development (i.e., expansion and deterioration) of ISR in concrete. It also highlights possible rehabilitation strategies and available management protocols.

8.2 PROGNOSIS AND THE POTENTIAL FOR FURTHER DETERIORATION

The prognosis of ISR-affected infrastructure should consider the likelihood of future deterioration and its associated structural implications. Therefore, laboratory test methods coupled with mathematical models are normally used for this purpose. Likewise, field monitoring (i.e., measurements of displacement, temperature, relative humidity, pressure, etc.) may also be effectively implemented as input or fitting parameters in mathematical models, supporting infrastructure owners in decision-making. The following sections

DOI: 10.1201/9781003188155-8

will present the most common laboratory techniques and engineering-based models to forecast the behaviour of ISR-affected concrete.

8.2.1 Laboratory techniques

Laboratory test procedures are normally the first approach used to evaluate the potential for future deterioration of ISR-affected concrete. Various test methods were developed over the last decades for this purpose, and although far from being perfect, they may provide valuable information. These tests are normally based on expansion or chemical procedures.

Expansion procedures

- **Objective:** Expansion procedures aim to estimate the potential of future expansion and deterioration of ISR-affected concrete. These procedures are generally divided into two categories: (a) assessing residual expansion or (b) absolute expansion. Although these tests do not represent the behaviour of affected structures or structural members in the field, they may provide valuable insights into ISR-induced development, such as the expected expansion kinetics (or rate) and its potential of levelling off over time (Merz & Leemann, 2013).
- **Sample preparation and methodology:** Expansion procedures are conducted following the coring of regions of interest of affected structures or structural members. Some common procedures should be performed to avoid variability in the test outcomes; first, the extracted cores from the distinct regions of interest should be carefully labelled and wrapped in plastic film to avoid moisture loss. Then, the wrapped cores should be kept in a controlled environment with a temperature of 23°C ± 2°C for five days. This allows for achieving consistent and homogeneous moisture content within the samples. Afterwards, the cores' ends are thoroughly prepared through cutting and grinding (or capping) processes until the desired length-to-diameter ratio of 1:2 (or close to) is achieved. Finally, studs are installed at the core ends to enable expansion monitoring over time. Figure 8.1 illustrates examples of cores' extraction (Figure 8.1a) labelling (Figure 8.1b), end studs' installation and storage (Figure 8.1c), and measurement (Figure 8.1d).

A number of test protocols have been developed worldwide for evaluating the expansion potential involving different test set-ups, environmental conditions, and measuring parameters (Saouma, 2021). Table 8.1 presents a comprehensive overview of current testing protocols, including set-up variations and primary outcomes for distinct ISR mechanisms. Among these, the most common procedure used in Canada, North America and various countries in Europe involves subjecting core samples to 38°C and 100% relative humidity for a duration of one year (or until the expansion levels off) to

Figure 8.1 (a) Cores' extraction, (b) Core's labelling, (c) End studs' installation and storage and (d) Core's measurement. Core sample preparation and measurement.

(Photos courtesy of Leandro Sanchez.)

determine the residual expansion potential (EXP) of the concrete. This procedure is normally used to assess AAR and AAR + DEF potential for further deterioration. Whether DEF is primarily intended to be appraised, the core specimens should be soaked in water at 38°C and monitored for one year (or until the expansion levels off) instead of stored under 100% RH (Kawabata et al., 2016; Martin, 2010; Martin et al., 2017).

Another approach commonly used during the appraisal of future deterioration caused by AAR entails the soaking of core specimens in 1M NaOH solutions at 38°C for one year; this procedure aims to measure the degree of absolute reactivity of aggregates (ABR). Figure 8.2 illustrates the test set-ups commonly used to obtain (EXP) and (ABR) outcomes.

Residual expansion tests are quite easy and relatively non-expensive procedures enabling the evaluation of the future expansion behaviour of ISR-affected concrete. However, important discrepancies are often observed when comparing the results obtained in the laboratory with the field performance of affected structures. Issues such as alkalis leaching, alkali release from aggregates, restraint/confinement, load and environmental conditions, which are different in the lab and field, have been pointed out as the main

Table 8.1 ISR residual expansion testing protocols

ISR mechanisms	Test set-up	Outcomes	Reference
AAR	95% RH and 38°C	Phases of residual expansion	Canadian Standards Association (CSA) (2000); ISE (1992); LCPC (2003); Multon et al. (2008)
AAR	1M NaOH and 38°C	Residual aggregate reactivity potential	Bérubé et al. (2002)
AAR	1M NaOH and 60°C	Residual expansion testing aggregates from the structure	Gao et al. (2011)
AAR	0.7M NaOH and 38°C, wrapped samples	Residual aggregate reactivity potential (improvement of 1M NaOH and 38°C method)	Zubaida (2020)
AAR	4% NaCl and 38°C	Structures exposed to salt	Swamy (1991, 1997)
AAR	1M NaOH and 80°C	Faster in reaching the ultimate residual expansion plateau	Bérubé et al. (2002)
DEF	Soaked in water and 38°C	Remaining potential of DEF-affected concrete	Martin et al. (2017)
DEF, DEF + ASR	Lime water and 23°C	Remaining potential of DEF-affected concrete	Folliard et al. (2006); Ramlochan et al. (2003)

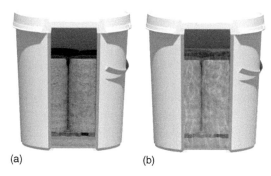

    (a)   (b)

Figure 8.2 Residual test set-ups performed to measure (a) the residual EXP and (b) the ABR. EXP can be used in both AAR and DEF cases, whereas ABR is used in AAR appraisal.

causes leading to the discrepancies verified (Bérubé et al., 2002; Sellier et al., 2009). Therefore, research is still required to improve the current test set-ups to better represent field performance. Nevertheless, more important than discussions on the reliability and efficiency of distinct residual expansion protocols is the interpretation and usage of the test outcomes obtained from them, which is a challenging task and should be conducted carefully and in a conservative fashion, accounting for the potential differences between lab and field.

Soluble alkalis

- **Objective**: Besides expansion tests, some chemical procedures may also estimate the potential of further expansion and deterioration of ISR-affected concrete; amongst them, the soluble alkalis method stands out as a promising protocol for appraising ISR future deterioration, particularly for AAR-affected concrete.

  It is widely known that the higher the amount of alkalis in the concrete pore solution (i.e., Na^+, K^+ and OH^-), the higher the likelihood of AAR-induced expansion and damage; therefore, assessing the remaining alkali content at a given age may provide insight into the remaining expansion yet to take place.

- **Sample preparation and methodology**: A few methods have been proposed over the past few decades to quantify the soluble alkalis in concrete; some of them include the extraction of concrete pore solution under pressure, while the vast majority consist of filtering under pressure of crushed and finely-ground concrete samples, followed by analysis of the alkali concentration in the obtained solution.

  Overall, the sample preparation process prior to filtering under pressure begins with the longitudinal splitting of concrete cores. The split core is then broken into large particles measuring approximately 25 mm. Subsequently, the sample undergoes multiple crushing stages until particle sizes of about 5 mm are reached. From this point, a representative subsample weighing 1 kg is selected and pulverized until the material reaches particle sizes of 150 µm. By splitting the material once more, two or three subsamples of 10 g are obtained and stored in airtight bags to prevent carbonation until testing.

The following techniques are considered the most reliable for extracting the pore solution and determining the pH and free alkali content: pore-water expression (PWE), in situ leaching (ISL), ex situ leaching (ESL), cold water extraction (CWE), hot water extraction (HWE) and espresso method (EM). Brief descriptions of these techniques are presented hereafter:

- **PWE**: This technique involves subjecting concrete samples to high pressure, resulting in pore solution extraction. PWE is a simple, effective and quite reliable test procedure widely implemented to evaluate the alkalis composition of concrete pore solution. PWE can be used for cement paste, mortar and concrete samples, yet attention should be paid to the interference of aggregates on the test outcomes, particularly for concrete specimens.

- **ISL:** This method is developed by the creation of a small hole in the concrete member under analysis, followed by the injection of a leaching solution; the concrete pore solution is then dissolved into the leaching solution, which is collected for further chemical analysis.
- **ESL:** ESL corresponds to the extraction and soaking of concrete cores into a leaching solution, allowing the alkalis to dissolve; the solution is then chemically evaluated and the alkalis amount quantified.
- **CWE:** CWE is a quite used technique due to its simplicity and minimal equipment requirements. Concrete powders (i.e., particles < 80 µm) are immersed in cold water (liquid-to-solid ratio equal to 1) for a 5 min leaching time. During this process, alkalis from the pore solution diffuse into water, which is then filtered and evaluated.
- **HWE:** This method follows the same general procedure of the CWE, yet concrete powders (i.e., particles < 160 µm) are boiled in 100 ml of deionized water over 10 min and left standing overnight, enhancing the diffusion of alkalis into the water.
- **EM:** This method involves the extraction of soluble alkalis by passing 300 ml of deionized boiling water through the concrete sample (i.e., 10 g of particles < 150 µm). Before being analysed, the solution is topped up to 500 mL with distilled water.

The literature shows that although all of the aforementioned methods display advantages and limitations, the EM is the technique that provides more accurate and less variable outcomes to assess the potential of further AAR-induced development (Plusquellec et al., 2017). However, there is still a debate in the research community on how to use those test outcomes to quantitatively estimate the potential of further AAR expansion in the field. On the other hand, the prognosis of other ISR mechanisms, such as DEF via pore solution analysis, is quite complex, and there is currently a lack of research, methods and quantitative data in this regard. However, it is accepted that the determination of pH, alkalis and sulphate contents of pore solution may help in understanding the potential of future deterioration caused by any ISR mechanism.

8.2.2 Prognosis estimation based on laboratory test procedures

Besides conducting tests that provide physical (i.e., expansion) or chemical (i.e., pore solution analysis) outcomes "individually", understanding how to combine and implement them into a prognosis protocol is probably amongst the most current challenges in the area. In this context,

Bérubé et al. proposed a framework for combining the information from laboratory tests into a protocol to estimate the potential expansion rate of AAR-affected structures in the field. Figure 8.3 illustrates the proposed framework (Bérubé et al., 2002), which is based upon research projects dealing with the assessment of AAR-affected dams.

In this framework, the residual and absolute expansions (EXP and ABR, respectively) are combined with soluble alkalis results (ALK) along with service conditions (i.e., temperature, relative humidity, confinement/restraint) to estimate the potential of further expansion (PFE) caused by AAR.

The coefficient EXP is normally the first coefficient gathered through residual expansion tests conducted at 38°C and 100% RH (see Section 8.2.1); this test is considered the most realistic laboratory test procedure to assess the PFE of AAR-affected members in service since the cores retrieved from affected structures are tested with their actual alkali

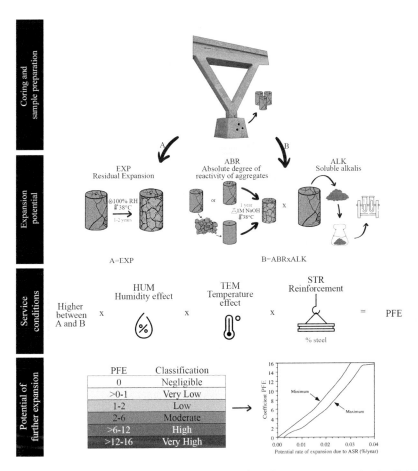

Figure 8.3 Framework for estimating AAR-induced expansion rate in the field.

Table 8.2 Classification of expansion potential, from Bérubé et al. (2002b)

Coefficient EXP – Residual concrete expansivity in the laboratory (core testing in air at > 95 RH and 38°C)

Exp./year %	Residual exp	EXP	Exp./year %	Residual exp	EXP
< 0.003	Negligible	0	0.015 to 0.02	Moderate	6
0.003 to 0.005	Very low	I	0.02 to 0.025	High	9
0.005 to 0.01	Low	2	0.025 to 0.03	High	12
0.01 to 0.015	Moderate	4	> 0.03	Very high	16

content (Bérubé et al., 1994). Annual residual expansion results ranging from negligible (i.e., less than 0.003% per year) to very high expansion (i.e., over 0.03% per year) may be obtained (Bérubé et al., 2002; Fournier et al., 2010) as per Table 8.2. However, it should be mentioned that EXP can often be underestimated due to alkalis leaching taking place over the test. Therefore, if the coefficient EXP is considered relatively low, it should be replaced by the product of the two coefficients ABR and ALK. The ABR coefficient is obtained following a one-year test at 38°C and 1M NaOH% (see Section 8.2.1), and the results can be classified from negligible (i.e., less than 0.04%) to very high (i.e., higher than 0.20%), as per Table 8.3. Otherwise, the soluble alkalis, and thus the coefficient ALK, is measured as per one of the procedures described in Section 8.2.1. Bérubé et al. proposed the use of the HWE method for such purpose (Bérubé et al., 1994). This index, which varies from very low (i.e., < 1.0 kg/m³) to very high (i.e., > 2.5 kg/m³), as illustrated in Table 8.4, indicates the remaining alkalis available to keep inducing further AAR expansion in the field. It is worth noting that

Table 8.3 Classification of the absolute degree of reactivity of the aggregates present in the concrete under study, based on expansion tests on cores in I N NaOH solution at 38°C or on CSA A23.2-14A-94 concrete prism tests on aggregates extracted from cores (ABR) (CSA, 2014)

I-year expansion (%) (After preconditioning in the case of cores)	Absolute degree of reactivity of aggregates[a]	Coefficient ABR[a]
< 0.04	Negligible	0 (or ≥ 0)
0.04 to 0.08	Low	I (or ≥ I)
0.08 to 0.12	Medium	2 (or ≥ 2)
0.12 to 0.20	High	3 (or ≥ 3)
> 0.20	Very high	4

[a] When testing cores, the qualification of the absolute degree of reactivity and the corresponding coefficient ABR are considered as minimum if the concrete specimens were abnormally fissured or porous compared to the overall concrete member under study or if the concrete is quite impermeable to the alkaline solution.

Table 8.4 Classification of the water-soluble alkali content of the concrete under study, as measured by the HWE method, after correction for aggregate contribution (ALK)

Corrected soluble alkali content (kg/m³ Na₂O$_{eq}$)	Classification	Coefficient ALK
< 1.0	Very low	0
1.0–1.5	Low	1
1.5–2.0	Medium	2
2.0–2.5	High	3
> 2.5	Very high	4

if, on the one hand, significant expansion obtained over the aforementioned expansion tests does not necessarily mean that the concrete under study will swell in service; for instance, if the humidity conditions in the field are too low to sustain alkali-silica reaction (ASR) and or the restraints a confinement effects are sufficiently high to suppress induced expansion. On the other hand, low expansions indicate that the concrete under study should not swell in service unless the concrete specimens tested are very deteriorated and or are much more porous than the concrete member under analysis. In other words, the samples used for testing do not represent, on average, the member appraised in the field. Once EXP, ABR and ALK are determined, parameters related to field conditions (i.e., humidity, temperature and stress state) should be obtained.

The humidity coefficient (HUM) can be gathered via commercial probes that can measure internal humidity along small drill holes (Bérubé et al., 1994); the range of values that can be obtained and their associated AAR risk are displayed in Table 8.5. If internal humidity values are not available, external humidity could be used to estimate the risk for AAR continuation as per Table 8.6.

The TEM is proposed to account for the effect of temperature on AAR-induced development; for example, the expansion tests conducted in the

Table 8.5 Classification of internal humidity conditions as regards the risk of ASR (HUM)

Relative humidity inside concrete (% RH)	"Humidity risk" for ASR	Coefficient HUM
< 80	Very low	0
80–85	Low	0.25
85–90	Medium	0.5
90–95	High	0.75
95–100	Very high	1

Table 8.6 Classification of external humidity conditions as regards the risk of ASR (HUM)

Humidity conditions (%RH)	Non-massive concrete[a]		Mass concrete[a]	
	"Humidity risk" for ASR	Coefficient HUM	"Humidity risk" for ASR	Coefficient HUM
Interior < 70%	Very low	0	Low	0.25
Interior 70%–80%	Low	0.25	Medium	0.5
Interior 80%–90%	Medium	0.5	High	0.75
Interior 90%–95%	High	0.75	Very high	1
Interior 95%–100% or immersed	Very high	1	'	1
Exterior not exposed to rain[b]	Medium	0.5	High	0.75
Exterior exposed to rain[b]	High	0.75	Very high	1
Exterior immersed or buried	Very high	1	'	1

[a] If the concrete component is at the same time exposed to humid and dry conditions, there is a risk for anisotropic expansion.
[b] In the case of a tempered climate as the one prevailing in North America.

laboratory are normally performed at 38°C, whereas most concrete structures exposed outdoors (e.g., North America) are submitted to yearly average lower temperatures. The idea of TEM is thus to take into account this effect, as per Table 8.7.

The coefficient of confinement/restraint (STR) is the most difficult parameter to estimate due to the often-limited amount of information (i.e., drawings, members detailing, etc.) of ageing infrastructure. The values proposed in Table 8.1 are based on "average results" encountered in the Institution of Structural Engineers report (ISE, 1992). It is important to note that the STR coefficient is valid in the direction parallel to the main rebars since literature

Table 8.7 Proposed values for the coefficient of thermal correction temperature coefficient (TEM)

Annual average temperature (°C)[a]	Coeff. of thermal correction TEM
< 0	0.4
0–10	0.55
10–20	0.7
20–30	0.85
> 30	1.0

a In most urban areas in Canada, the annual average temperature is between 0°C and 10°C.

Table 8.8 Proposed values for the coefficient of correction for reinforcement and stresses applied to concrete in service (STR)

Amount of steel reinforcement (%)[a]	Coeff. of correction for stresses STR	Internal or external restraints (MPa)[b]	Coeff. of correction for stresses STR
0	1.0	0	1.0
0.25	0.75	0.25	0.85
0.5	0.55	0.5	0.7
0.75	0.4	0.75	0.55
1	0.3	1	0.4
2	0.25	1.5	0.3
≥ 3	0.2	2	0.2
		≥ 3	0.1

[a] When rebars are installed on a single plane (1D or 2D reinforcement) or on many parallel planes (2D or 3D reinforcement) but without any anchorage between the different planes, the coefficient STR applies in the direction(s) of the rebars only.
[b] In the cases of uniaxial (1D) or biaxial (2D) compressive stresses, the coefficient STR applies in the direction of the stresses only.

shows that due to anisotropy, induced expansion is redistributed and becomes higher in directions perpendicular to the main reinforcement bars.

Finally, upon obtaining all parameters, the potential of further expansion (PFE) of the structure or structural member under evaluation can be computed and correlated with a potential expansion rate(% per year) as per Figure 8.3.

8.2.3 Modelling

Mathematical models play a crucial role in the representation and understanding of physicochemical phenomena. A wide range of models have been developed in the past to describe ISR-induced expansion and deterioration, from describing the chemical phenomena to evaluating the performance and capacity of critical infrastructure in the field. The goal of this chapter is not to debate the pros and cons of each of those methods. Moreover, it is out of the scope of the current book to discuss "structural" models, normally run by finite elements, and to appraise service and ultimate limit states of affected structures. This is considered the following step of our proposed condition assessment protocol, normally conducted by structural engineers after diagnosis and prognosis tasks. Yet, the idea of this section is to discuss models that aim to evaluate the potential of further continuation of ISR-induced development and ultimately estimate the expansion as a function of time of the affected concrete in the "material's scale".

ISR models can be divided into four categories: micromodels (based on ion diffusion/reaction products), micro-mesomodels (based on secondary products generation), mesomodels (based on internal pressure) and macromodels (based on expansion). Micro-, micro-meso- and mesomodels are often considered for materials' scale appraisal, whereas macromodels can be used for material and structural evaluations (Esposito, 2015).

One example of a very promising micro-mesomodel to describe AAR effects in concrete is a chemo-mechanical elastic model developed by researchers from the University Paul Sabatier in Toulouse, France (Multon et al., 2009). This model considers the chemical interactions between the alkalis in the concrete pore solution and the non-stable siliceous phases from the aggregates within a representative elemental volume (REV) of concrete; parameters such as alkali diffusion into aggregates, secondary products formation and deposition in the concrete porosity are considered. As a result, a volumetric amount of AAR secondary products is formed and induces expansion, as illustrated in Figure 8.4; moreover, these products, after filling the initial porosity of the concrete, produce pressure which induces cracks in the cement paste once the tensile stresses generated exceed the tensile strength of the concrete.

The initial calibration of this chemo-mechanical elastic model was based on experimental data obtained in mortars, which yielded satisfactory results. The model was further validated using extensive data on crack generation

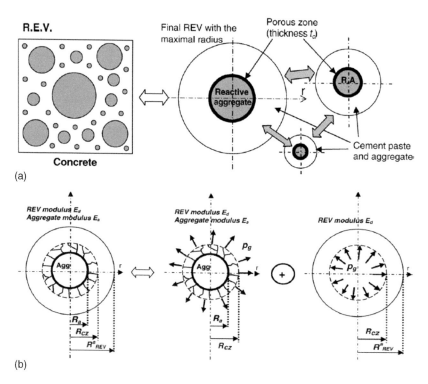

Figure 8.4 Micro-mesomodel to describe AAR-induced product formation (a) and stresses and cracking generated (b).

and propagation and mechanical properties reductions from Sanchez et al. (Sanchez et al., 2014), demonstrating its promising character to predict induced expansion and damage. Another example of an interesting AAR mesoscopic model is the one proposed by Dunant and Scrivener at EPFL in Lausanne, Switzerland (Dunant & Scrivener, 2010). This approach uses a finite element framework, where the aggregates are assumed to be completely spherical, both the aggregates and cement paste are considered quasi-brittle, and the secondary products generated behave in a linear-elastic fashion. The model considers that AAR randomly generates secondary products that fill voids, flaws or previous cracks within the reactive aggregates, tending to form "pockets" within the particles. An enrichment function (Moës et al., 2003) is used to simulate the exact contact between the secondary products and the aggregates. Induced expansion is then developed whenever the amount of products overcomes the initial porosity of the aggregates. Furthermore, cracks and reduced stiffness are computed once expansion is induced. As for the chemo-mechanical elastic model, encouraging correlations were observed between simulations and experimental data (Haha et al., 2007) concerning crack propagation and stiffness reduction (Dunant & Scrivener, 2010).

Macromodels, primarily aiming to assess induced expansion, which account indirectly for the distinct aspects of ISR physicochemical processes, are probably the most used type of models adopted for engineering and decision-making purposes. Most of these models are based on Larive's model (Larive, 1997), which is a semi-empirical approach that was developed to appraise ASR-induced expansion after a large experimental campaign in the laboratory. The model describes induced expansion and ASR kinetics using three parameters: latency (τ_l), characteristic (τ_c) and ultimate expansion (ε^∞), as shown in Equation 8.1.

$$\varepsilon(t,\theta) = \frac{1 - e^{\frac{-t}{\tau_c(\theta)}}}{1 + e^{\frac{(t-\tau_l(\theta))}{\tau_c(\theta)}}} * \varepsilon^\infty \tag{8.1}$$

The induced expansion described by Larive's model often presents an S-shape (i.e., sigmodal) curve, which may be broken down into four phases, as illustrated in Figure 8.5. Initially, in phase A, the curve represents the formation and accommodation of ASR secondary products within the reactive aggregate particles and surrounding flaws in the cement paste (i.e., ITZ), resulting in little to no expansion. The ascending period (phase B) of the curve corresponds then to the swelling of ASR secondary products upon moisture uptake from the surroundings; during this phase, significant expansion and crack formation are expected, explaining the fast kinetics and convex shape of the curve. The subsequent phase (phase C), distinguished from the previous by a concave behaviour, suggests the decrease in expansion rate due to the increased number of cracks generated in the system, creating

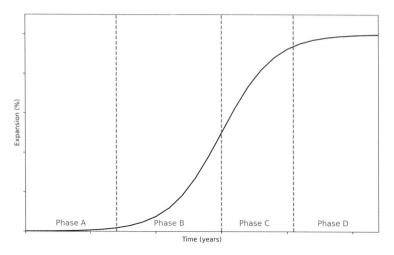

Figure 8.5 S-shape curve described by Larive's model for ASR-induced development.

a supplementary space to accommodate secondary products. In the final phase, or phase D, the reactants (i.e., alkalis and/or silica) are expected to be depleted from the system, which makes the curve level off.

Larive's model has shown to be a quite promising approach to describe ASR-induced development and has been successfully used in numerous applications; however, since latency (τ_l) and characteristic (τ_c) parameters are primarily mathematical terms rather than parameters representing physical or chemical variables; they may be hardly estimated and are often "fit" in a case-by-case approach to match the behaviour observed by the affected concrete under evaluation. This aspect limits its ability to predict behaviour of ASR-affected concrete in the field whether no monitoring or tests are conducted.

Trying to better understand and represent the influence of key components of ASR on latency (τ_l) and characteristic (τ_c) parameters, De Grazia et al. (2021) proposed that measurable parameters, such as temperature, aggregate type (i.e., fine or coarse) and reactivity (i.e., marginal, moderate, high and very high), humidity and alkali content be used to "calibrate" and or modify latency (τ_l) and characteristic (τ_c) parameters. The authors suggested that new coefficients should be added in the initial Larive's equation for this purpose; these coefficients were established in the laboratory using a wide variety of test procedures, mix proportions and aggregate reactivity levels. The modified form of Larive's model considering the new coefficients is presented in Equation 8.2.

$$\varepsilon(t,\theta) = \varepsilon(t)\varepsilon^{\infty} = \frac{1 - e^{-\frac{t}{\tau_c k_{c,T} k_{c,RH} k_{c,\%A} k_{c,E}}}}{1 + e^{-\frac{(t - \tau_l k_{L,T} k_{L,RH} k_{L,\%A} k_{L,E})}{\tau_c k_{c,T} k_{c,RH} k_{c,\%A} k_{c,E}}}} * \left(k_{Inf,T}\, k_{Inf,RH}\, k_{Inf,E}\, k_{Inf,\%A}\right)\varepsilon^{\infty},$$

$$(8.2)$$

where t is the elapsed time; $\varepsilon(t)$ is the expansion at a given elapsed time; ε^∞ is the maximum expansion at infinity (or ultimate expansion); τ_c is the characteristic time (as a function of the aggregate type and nature/reactivity); $k_{c,T}$, $k_{c,RH}$, $k_{c,\%A}$, $k_{c,E}$ is the temperature, humidity, alkali content and exposure coefficients impacting the characteristic time; $k_{c,T}k_{c,RH}k_{c,\%A}k_{c,E}$ is the temperature, humidity, alkali content and exposure coefficients impacting the latency time; $k_{Inf,T} k_{Inf,RH} k_{Inf,E} k_{Inf,\%A}$ is the temperature, humidity, exposure and alkali content coefficients influencing the maximum expansion.

The model proposed by De Grazia et al. (2021) showed promising results in describing the behaviour of ASR and ACR-induced expansion in the laboratory, and it has even been adopted and slightly modified to appraise an ASR-affected reinforced concrete overpass after nearly 50 years of service in Quebec City, Canada (Gorga et al., 2018). Yet, it is worth noting that De Grazia's model has been primarily developed to estimate the behaviour of AAR-induced development in the laboratory, and thus care should be taken for its use in field applications.

Another contribution to the use of semi-empirical macromodels to describe AAR-induced expansion has been made by Nguyen et al. (2022). In this model, the conventional Larive's model is adopted, and its latency (τ_l) and characteristic (τ_c) parameters are calibrated by laboratory tests (i.e., concrete prism test – CPT) and then used to estimate the behaviour of concrete exposed to field conditions but bearing the same raw materials and mix proportions to the mixture tested in the laboratory. To improve the correlation between laboratory and field, this model accounts for (a) the amount of alkalis leached over the accelerated test in the laboratory through the computation of a "virtual expansion" showing no-leaching and (b) the potential alkalis contribution from the aggregates. Figure 8.6 illustrates the flowchart proposed by Nguyen et al. (Nguyen et al., 2022) with the concepts

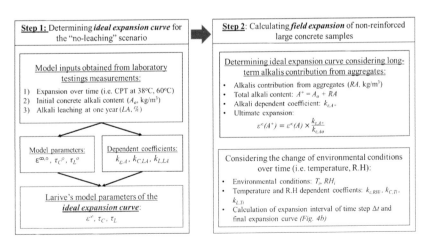

Figure 8.6 Flowchart proposed for the semi-empirical approach as per (Nguyen et al., 2022).

of the proposed approach. Strong correlations to estimate field behaviour from laboratory testing were obtained, showing a promising character of the proposed model (Nguyen et al., 2022).

Although the number of models developed for AAR applications is probably much higher than for other ISR mechanisms, numerical works have also been conducted to describe induced expansion caused by internal sulphate attack, particularly DEF. As for AAR, most of the approaches use Larive's model to describe the expansion development over time. In this context, an interesting approach has been proposed by Brunetaud (Brunetaud, 2005), where a new time-dependent parameter was proposed to Larive's model to better describe the long-term behaviour of DEF in concrete. The proposed model is presented in Equation 8.3, and it has shown great results in describing DEF-induced expansion in the laboratory.

$$\varepsilon\left(t\right) = \varepsilon_{\infty} \frac{1 - e^{\left(\frac{t}{\tau_c}\right)}}{1 + e^{\left(\frac{t}{\tau_c} + \frac{\tau l}{\tau_c}\right)}} * \left(1 - \frac{\Phi}{\delta + 1}\right),$$

(8.3)

where Φ and δ are corrective parameters, and t is elapsed time (Kawabata et al., 2016).

As presented in Chapter 2, AAR and DEF are often combined in the field due to their synergetic character. Therefore, developing a model that couples both phenomena is crucial. Martin (Martin, 2010) proposed a novel semi-empirical approach for coupled AAR and DEF mechanisms, which is based upon Larive's (1997) and Brunetaud's (2005) approaches for AAR and DEF, respectively. The proposed model (presented in Equation 8.4) accounts for the contribution of both AAR and DEF separately, yielding a total expansion over time for each mechanism.

$$\varepsilon\left(t\right) = \varepsilon_{DEF} + \varepsilon_{ASR} = \varepsilon_{\infty\_DEF} * \frac{1 - e^{\left(-\frac{t}{\tau_{CDEF}}\right)}}{1 + e^{\left(\frac{t - \tau_{LDEF}}{\tau_{CDEF}}\right)}}$$

$$\times * \left(1 - \frac{\Phi}{\delta + 1}\right) + \varepsilon_{\infty\_ASR} * \times \frac{1 - e^{\left(-\frac{t}{\tau_{C\_ASR}}\right)}}{1 + e^{\left(\frac{t - \tau_{L\_ASR}}{\tau_{C\_ASR}}\right)}}$$

(8.4)

Martin et al. (2010) successfully applied this combined model to represent experimental data obtained from specimens distressed by the combined effects of ASR and DEF at 100% RH, effectively capturing the expansion over time caused by each mechanism, as per Figure 8.7.

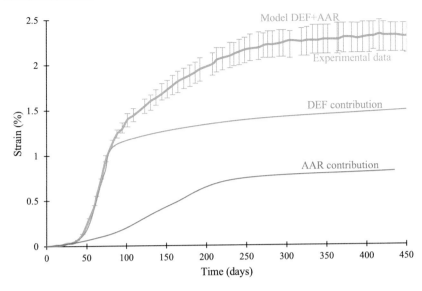

Figure 8.7 Experimental versus numerical results.

(Adapted from R. P. Martin et al., 2010 approach.)

8.3 MANAGEMENT AND REHABILITATION OF ISR-AFFECTED CONCRETE INFRASTRUCTURE

Upon identifying the current cause(s) and extent of damage, as well as estimating future deterioration and likely structural implications of ISR-affected infrastructure, efficient management protocols should be implemented to better cope with the deterioration as a function of time. Normally, management protocols comprehend two major steps: (a) a series of field and laboratory test procedures along with modelling, to appraise and ensure serviceability and safety as a function of time, and; (b) rehabilitation strategies aiming to stop or at least mitigate the rate of deterioration over time. The next two sections will cover the most common management protocols and rehabilitation procedures implemented in ISR-affected structures.

8.3.1 Management protocols

From previous chapters' discussions, it is quite clear that management protocols for ISR-affected infrastructure should bear at least four distinct yet complementary steps: (a) field assessment, where visual and non-destructive techniques are used to gather preliminary information; (b) laboratory tests, where chemical, microscopic and mechanical tests are conducted to understand the cause(s) and extent of deterioration (i.e., diagnosis); (c) data analysis, where the prognosis appraisal along with the current and future structural implications are assessed; and (d) decision-making, where maintenance and rehabilitation strategies are decided. Figure 8.8 illustrates such a hypothetical four-phase management protocol.

Over time, a few specific management protocols have been developed to help infrastructure owners better cope with ISR-affected infrastructure. Amongst them, four protocols stand out as promising frameworks, starting from literature survey and condition assessment to rehabilitation strategies: (a) Bérubé et al. (Bérubé & Fournier, 2005), (b) RILEM (Godart et al., 2013), (c) (Fournier et al., 2010) and (d) IFSTTAR (LCPC, 2003). Figure 8.9 illustrates the content of the proposed flowcharts.

From the aforementioned flowcharts, one observes that the proposed management protocols bear similarities and differences, yet the four phases illustrated in the hypothetical ISR-management protocol appear (with more or less detail) in each of them. Moreover, these protocols incorporate several of the field and laboratory test procedures (i.e., visual inspection, cracking monitoring, non-destructive testing, microscopy and mechanical tests) discussed in detail in Chapters 3, 4, 5 and 6. Prognosis is also discussed on them, based on laboratory tests and modelling (Section 8.2 of this chapter). However, they are rather descriptive and thus subjective by nature, heavily relying on the expertise of the engineer(s) dealing with the affected structure to decide whether further assessment is required following visual inspection. Nevertheless, some descriptive and

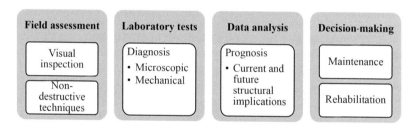

Figure 8.8 Comprehensive management protocol for ISR-affected structures.

	Bérubé et al.	RILEM	FHWA	IFSTTAR
Field assessment (VI, NDTs)	✓	✓	✓	✓
Laboratory tests for diagnosis (petrography, SDT, DRI, SEM)	✓	✓	✓	✓
Data analysis for prognosis (RE, soluble alkalis)	✓	✗	✓	✓
Decision-making (maintenance and rehabilitation strategies)	✓	✗	✗	✓

Figure 8.9 Comparison among the proposed management protocols: Bérubé et al., RILEM, FHWA and IFSTTAR.

semi-quantitative guidelines are provided to better guide infrastructure owners in decision-making, particularly regarding surface deterioration signs, such as the classification of the probability of AAR (based on expansion/displacement of members and surface deterioration), the classification of the AAR development degree and the classification of the importance of crack widths and cracking index values, as illustrated in Tables 8.9 to 8.11, respectively.

Regardless of the protocol used, if the engineer(s) in charge of the evaluation considers that further assessment is required, coring is then conducted and laboratory (i.e., chemical, microscopic, mechanical, etc.) tests are performed with the aim of diagnosing the current deterioration (i.e., cause and extent of damage). For this step, various approaches are available in the literature, yet a broad discussion, along with the implementation of the multi-level assessment, considered a quite reliable protocol, is presented in Chapter 7 of this book. Upon diagnosis completion, the prognosis is required and then the evaluation of the potential of further deterioration may be achieved via laboratory tests (e.g., residual expansion, soluble alkalis), modelling or a

Table 8.9 Classification of probability of AAR occurrence

| Feature | Probability of AAR occurrence | | |
	Low	Medium	High
Expansion and/or displacement of elements	None	Some evidence (e.g., closure of joints in pavements, Jersey barriers, spalls, misalignments between structural members)	Fair to extensive signs of volume increase leading to spalling at joints, displacement and/or misalignment of structural members
Cracking and crack pattern	None	Some crack patterns typical of ASR (e.g., map cracking or cracks aligned with major reinforcement or stress)	Extensive map cracking or cracking aligned with major stress or reinforcement
Surface discoloration	None	Slight surface discoloration associated with some cracks	Many cracks with dark discoloration and adjacent zones of light-coloured concrete
Exudations	None	White exudation around some cracks, possibility of colourless, jelly-like exudations	Colourless, jelly-like exudations readily identifiable as ASR gel associated with several cracks

Source: Modified from (CSA A864-00 (R2005), 2005)

Table 8.10 Classification of AAR development degree (CSA A864-00 (R2005), 2005)

AAR development degree	Nature extent of features
Low	No gel present, no sites of expansive reaction, presence of other indicative features were rarely found. Small amount of isolated ASR gel was sporadically present in the cement paste, but very few reactive particles were identified. Absence of other expansive-related products and features caused by some other mechanisms. Quality of concrete microscopically is reasonably good.
Medium to High	Presence of some or all features is generally consistent with AAR, such as: • cracking and microcracking, especially when associated with known reactive aggregates. • presence of potentially reactive aggregates. • internal fracturing of known reactive aggregates. • darkening of cement paste around aggregate particles, cracks or voids. • presence of reaction rims around the internal periphery of reactive aggregate particles. • presence of damp patches on core surfaces.
Very High	Presence of features such as evidence of sites of expansive reaction, that is, locations within the concrete where evidence of reaction and emanation of swelling pressure can be positively identified, for example, streaming of ASR gel from a reacted and cracked aggregate particle into the adjoining cement paste with development of cracks both in the cement paste and along the paste-aggregate interface. Presence of ASR gel in cracks and voids associated with reactive particles and readily visible to normal or corrected-to-normal vision or under low magnification.

Table 8.11 Classification of the importance of crack widths (CSA, 2014) and cracking index values (Fasseu & Michel, 1997)

CSA			IFSTTAR	
Crack width (mm)	Description		CI	Cracking importance
< 0.1	Fine	Usually present	0–0.5	Negligible
0.1–0.3	Normal	Normal limit for RC	0.5–1	Marginal
0.3–0.5	Large	Over limit	1–2	Moderate
0.5–1.0	Moderately wide	Record all	2–5	High
2.0–1.0	Wide	Refer to Engineer	5–10	Very high
>5.0–10.0	Very wide	Refer to Engineer	> 10	Ultra-high

combination of both (Chapter 8). Structural evaluation should also be conducted by experts at this phase, taking into consideration both current and future deterioration based on diagnosis and prognosis, respectively. The acceptable threshold of induced expansion and associated deterioration depends on the type and importance of the affected structure; however, it is quite common to define the "end of service life" of structures affected by ISR, when its members reach, expansion levels of 0.20%, since such a value indicates yielding of the rebars and thus potential structural implications. It is worth noting that this threshold value may be selected on a case-by-case basis and could be much lower whether serviceability aspects are considered. Finally, depending on the current damage degree and the potential for further deterioration, infrastructure owners or engineers in charge of critical infrastructure may select rehabilitation strategies, which may largely vary and depend upon the structure type, importance and deterioration degree. The next section discusses some of the techniques used to mitigate/reduce deterioration caused by ISR in concrete infrastructure.

8.3.2 Rehabilitation techniques

The efficiency of rehabilitation strategies depends on the ISR type, along with the current damage and the potential for further deterioration. Moreover, it is quite clear that the sooner maintenance and rehabilitation strategies are implemented, the more efficient they are and the less structural implications they might have. Figure 8.10 associates tasks implemented in ISR-affected

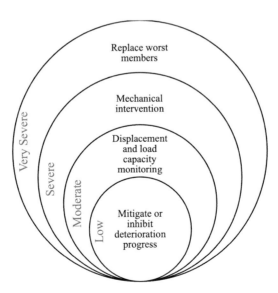

Figure 8.10 Classification of residual expansion potential and its correspondent intervention strategy.

concrete structures as a function of their degree of damage (i.e., low, moderate, severe and very severe). The following subsections discuss the strategies adopted in each of these deterioration levels.

8.3.2.1 Mitigate or inhibit deterioration progress

The strategies employed to mitigate or inhibit the progress of deterioration in concrete infrastructure usually involve limiting the sustaining conditions of the main damage mechanism(s) inducing deterioration. As discussed in Chapter 2, water has been identified as a key factor in ISR, especially in AAR and DEF cases. Therefore, reducing water in the system can effectively mitigate (i.e., decrease the deterioration rate) of affected concrete (De Souza and Sanchez, 2023). Common strategies in this regard include drainage improvement, coatings and crack injections. However, when inhibiting the deterioration progress is the goal, chemical aspects should also be considered, and thus specific compounds may be used, such as lithium-based products.

8.3.2.1.1 Drainage system

Structural members that are constantly exposed to the environment and water, such as highway structures (i.e., abutments, piers, girders, pile caps, decks, curbs and sidewalks), foundation blocks and dams, require an efficient drainage system. The implementation of a well-designed drainage system allows for the reduction of internal humidity, aiming to decrease the expansion rate of ISR-affected concrete. Therefore, it is crucial to conduct regular inspections and cleaning of drainage systems along with performing necessary maintenance to ensure their optimal operation. In the case of ISR-affected structures, drainage is often combined with cladding, which provides an additional layer of protection.

8.3.2.1.2 Moisture control by coatings

The implementation of methods that efficiently reduce the internal relative humidity of concrete is essential to control ISR-induced development. While surface coatings that prevent moisture ingress are advantageous, it is also important to consider the moisture within the concrete to mitigate AAR. Thus, when considering surface coating, the goal is not to limit external water infiltration but rather to facilitate internal humidity reduction. This promotes the gradual drying of the concrete and leads to an overall decrease in internal relative humidity. Coatings, sealers and waterproofs used for moisture control of ISR-affected concrete should have specific properties, including resistance to water absorption, the ability to penetrate into measurable depths, durability against ultraviolet (UV) degradation and long-term stability in alkaline environments.

The efficiency of products such as silanes, siloxanes, and polyurethanes in controlling moisture content and minimizing AAR-induced expansion has been thoroughly appraised and attested in numerous studies (Bérubé et al., 2002a; Champagne et al., 2019). These products are also recognized for their ability to enhance frost resistance, reduce chloride ingress and protect steel reinforcement from corrosion (Deschenes, 2017).

8.3.2.1.3 Crack injection

Crack filling (or injection) is considered in cases where ISR-induced cracking is excessively present on the surface of affected concrete. Thresholds for crack filling have been established at 0.15 mm and 0.30 mm for reinforced and non-reinforced concrete members, respectively, while for prestressed concrete, the limit is lower (i.e., 0.10 mm) (Fournier et al., 2010). Crack filling involves the use of flexible materials, such as flexible grouts, caulkings and polymers, preventing water and other substances from entering through the cracks. While rigid polymers and cement grouts can be used to temporarily stabilize cracks, cracks adjacent to the filled ones may eventually appear due to the rigid nature of these materials and their strong bond to the concrete substrate. Crack injection is a symptomatic treatment and should be used to restore concrete monolithic conditions rather than mitigating ISR; it is therefore recommended that crack filling be used as part of a comprehensive approach to inhibit induced deterioration, such as combining it with surface coatings or sealers (LCPC, 2003).

8.3.2.1.4 Lithium-based admixtures

Lithium-based admixtures (whether applied topically or introduced into the concrete through vacuum impregnation or electrochemical chloride removal process) have been widely used in both new and existing concrete structures to mitigate ASR-induced expansion and slow down its damage development (Folliard et al., 2003; Stokes et al., 1997; E.R. Giannini, et al. 2011). Among these, lithium nitrate solutions, being pH neutral, easy to handle and possessing higher penetration rates than lithium hydroxide solutions, have gained popularity in recent years.

> **Topical treatment:** Lithium may be topically applied to ASR-affected concrete members; however, previous works of topical lithium application have raised concerns about the extent of penetration and its ability to reduce ASR deterioration. It is important to note that the extent of concrete distress at the time of the treatment significantly influences the potential for lithium ingress. This means that the sooner the lithium is applied, the better and more efficient it may be. Results indicate that for concrete cracks with average widths of 0.2 mm or more, lithium penetration can reach depths of about 50

mm (Ekolu et al., 2017; Kawamura & Kodera, 2005; Kobayashi & Takagi, 2020; Souza et al., 2017).

Electrochemical impregnation method: Electrochemical methods can be used to impregnate lithium into concrete. This technique has been widely employed with various lithium compounds, such as lithium borate, -nitrate and -hydroxide. While only a limited number of tests have been conducted to determine the extent of lithium penetration through electrochemical impregnation, it has been observed that a penetration depth of at least 30 mm is often achieved (De Souza, 2016).

Vacuum impregnation: Vacuum impregnation creates a negative pressure that enables repair products, including lithium-based admixtures, to penetrate into materials, filling and interconnecting cracks, voids and even microcracks. Lithium has been applied via vacuum to treat ASR-induced deterioration in a wide range of concrete members, such as abutment walls, sidewalks, parapets and decks (Fournier et al., 2010).

8.3.2.2 Displacement monitoring and loading carry capacity

The ongoing swelling deterioration process in ISR-affected structures can induce movement, potentially leading to structural instability. In such a sense, monitoring the structure displacement through devices and sensors such as inclinometers, extensometers, accelerometers and fibre optics may indicate the intervention time and structure member to be intervened. Furthermore, monitoring the structure loading carrying capacity with active solutions (e.g., distributed load balancing, active dumping systems or active vibration control) can prevent further damage to the structure.

8.3.2.3 Mechanical intervention

When the damage extent reaches a point where it may cause serious structural implications or when other mitigating systems cannot be applied, mechanical intervention becomes necessary.

Strengthening: Post-tensioning (using tendons or cables) is currently the primary structural-level intervention employed to restore the integrity of ISR-affected infrastructure; it is commonly used for non-massive structural reinforced concrete members using 1D or 2D encapsulation (CSA, 2000). Besides post-tensioning, the introduction of reinforcement in the form of straps, steel plates and bolts tensioning has proven to be efficient in containing ASR-induced expansion (ISE, 1992). Moreover, rock anchors and encapsulation methods have been utilized to restrain expansion and movement in mass concrete foundations, including tower blocks (Villemure et al., 2019).

Stress relief via slot cutting: Stress relief via slot cutting is one of the common strategies to relieve stresses at selected locations of ISR-affected structures and thus isolate further swelling (ISE, 1992). This technique is widely used in massive structures, particularly in dams (Charlwood & Solymar, 1995). However, it is important to note that slot cutting may only provide a "temporary solution" since ISR (e.g., AAR or DEF) are ongoing deterioration processes that keep evolving; therefore, further cuts may be necessary over time, increasing the cost of the rehabilitation process. Slot cutting often causes the "local" increase in expansion rate before the relief gap is closed (Charlwood & Solymar, 1995). Moreover, since the cut alters the overall stress state of surrounding areas, additional reinforcement may be necessary to ensure the stability of concrete elements during and after slot cutting.

8.3.2.4 Replacement

While the replacement of ISR-affected concrete may be the most efficient remedial measure, it is rarely economically feasible. A common approach used in practice is to replace only a small portion of the affected structure while modifying or reinforcing the vast majority of deteriorated structural components so that the structure meets acceptable service conditions (Blight & Ballim, 2000).

REFERENCES

Bérubé, M.-A., Chouinard, D., Pigeon, M., Frenette, J., Rivest, M., & Vézina, D. (2002a). Effectiveness of sealers in counteracting alkali-silica reaction in highway median barriers exposed to wetting and drying, freezing and thawing, and deicing salt. *Canadian Journal of Civil Engineering*, 29(2), 329–337. https://doi.org/10.1139/l02-010

Bérubé, M.-A., Duchesne, J., Dorion, J. F., & Rivest, M. (2002b). Laboratory assessment of alkali contribution by aggregates to concrete and application to concrete structures affected by alkali-silica reactivity. *Cement and Concrete Research*, 32(8), 1215–1227. https://doi.org/10.1016/S0008-8846(02)00766-4

Bérubé, M. A., & Fournier, B. (2005). *Outil d'évaluation et de gestion des ouvrages d'art affectés de réactions alcalis-silice (RAS)*.

Bérubé, M.-A., Frenette, J., Pedneault, A., & Rivest, M. (2002). Laboratory assessment of the potential rate of ASR expansion of field concrete. *Cement, Concrete and Aggregates*, 24(1), 13. https://doi.org/10.1520/CCA10486J

Bérubé, M. A., Pedneault, J. F., & Rivest, M. (1994). *Laboratory assessment of potential for future expansion and deterioration of concrete affected by ASR*.

Blight, G. E., & Ballim, Y. (2000). Properties of AAR-affected concrete studied over 20 years. In M. A. Bérubé, B. Fournier, & B. Durand (Eds.), *11th international conference on Alkali-aggregate reaction* (pp. 1109–1118). Centre de recherche interuniversitaire sur le béton (CRIB).

Brunetaud, X. (2005). *Étude de l'influence de différents paramètres et de leurs interactions sur la cinétique de l'amplitude de la réaction sulfatique interne au béton* [PhD]. Châtenay-Malabry, Ecole centrale de Paris.

Canadian Standards Association (CSA). (2000). *Guide to the Evaluation and Management of Concrete Structures Affected by Alkali-Aggregate Reaction*.

Champagne, M., Roy-Tremblay, M., Fournier, F., Duchesne, F., & Bissonnette, B. (2019). Long-term effectiveness of sealers in counteracting alkali-silica reaction in highway median barriers exposed to wetting and drying, freezing and thawing, and de-icing salts. *17th Euroseminar on Microscopy Applied to Building Materials.*

Charlwood, R. G., & Solymar, Z. V. (1995). Long-term Management of AAR-Affected Structures – An International Perspective. *AAR in Hydroelectric Plants and Dams: Proceedings of the 2nd International Conference*, 19–55.

CSA. (2000). *CSA A864-00 - Guide to the Evaluation and Management of Concrete Structures Affected by Alkali-Aggregate Reaction.* Canadian Standards Association.

CSA. (2014). *A23.1-14/A23.2-14 Concrete materials and methods of concrete construction/Test methods and standard practices for concrete.* Canadian Standards Association (CSA).

CSA A864-00 (R2005). (2005). Guide to the Evaluation and Management of Concrete Structures Affected by Alkali-Aggregate Reaction. In *CSA A864-00 R2005 Constr. Mater. Build. Struct. Build. Concr. Struct* (p. 108).

De Grazia, M. T., Goshayeshi, N., Gorga, R., Sanchez, L. F. M., Santos, A. C., & Souza, D. J. (2021). Comprehensive semi-empirical approach to describe alkali aggregate reaction (AAR) induced expansion in the laboratory. *Journal of Building Engineering*, 40(January). https://doi.org/10.1016/j.jobe.2021.102298

De Souza, D. J., & Sanchez, L. F. M. (2023). Evaluating the efficiency of SCMs to avoid or mitigate ASR-induced expansion and deterioration through a multi-level assessment. *Cement and Concrete Research*, 173, 107262. https://doi.org/10.1016/j.cemconres.2023.107262

De Souza, L. M. S. (2016). *Electrochemical lithium migration to mitigate alkali-silica reaction in existing concrete structures.* Delft University of Technology.

Deschenes, R. A. (2017). *Mitigation and Evaluation of Alkali-Silica Reaction (ASR) and Freezing and Thawing in Concrete Transportation Structures* [PhD]. University of Arkansas.

Dunant, C. F., & Scrivener, K. L. (2010). Micro-mechanical modelling of alkali-silica-reaction-induced degradation using the AMIE framework. *Cement and Concrete Research*, 40(4), 517–525. https://doi.org/10.1016/j.cemconres.2009.07.024

Ekolu, S., Rakgosi, G., & Hooton, D. (2017). Long-term mitigating effect of lithium nitrate on delayed ettringite formation and ASR in concrete – Microscopic analysis. *Materials Characterization*, 133, 165–175. https://doi.org/10.1016/j.matchar.2017.09.025

Esposito, R. (2015). *The deteriorating impact of Alkali-Silica reaction on concrete: Expansion and mechanical properties.* Delft University of Technology.

Fasseu, P., & Michel, M. (1997). *Détermination de l'indice de fissuration d'un parement de béton; Méthode d'essai LCPC N0. 47.*

Folliard, K. J., Barborak, R., Drimalas, T., Du, L., Garber, S., Ideker, J., Ley, T., Williams, S., Juenger, M., Fournier, B., & Thomas, M. D. A. (2006). *Preventing ASR/DEF in new concrete: Final report (0-4085-5).* Texas Department of Transportation and the Federal Highway Administration.

Folliard, K.J., Thomas, M.D.A., Ideker, J.H., East, B. & Fournier, B. (2009). Case Studies Treating ASR-Affected Structures with Lithium Nitrate. *Transportation Research Board Annual Meeting 2009 Paper #09-2685.*

Fournier, B., Bérubé, M. A., Folliard, K., & Thomas, M. (2010). *Report on the diagnosis, prognosis, and mitigation of Alkali-Silica Reaction (ASR) in transportation structures.*

Gao, X. X., Multon, S., Cyr, M., & Sellier, A. (2011). Optimising an expansion test for the assessment of alkali-silica reaction in concrete structures. *Materials and Structures*, 44(9), 1641–1653. https://doi.org/10.1617/s11527-011-9724-y

Giannini, E.R., Bentivegna, A.F., & Folliard, K.J. (2011). Coatings and overlays for concrete affected by Alkali-Silica reaction. In V. Mechtcherine, U. Schneck, & M. Grantham (Eds.), *4th international conference on concrete repair*. Dresden, Germany.

Godart, B., Rooij, M., & Wood, J. G. M. (2013). *Guide to diagnosis and appraisal of AAR damage to concrete in structures: Part 1 diagnosis (AAR 6.1)* (RILEM). Springer.

Gorga, R. V., Sanchez, L. F. M., & Martín-Pérez, B. (2018). FE approach to perform the condition assessment of a concrete overpass damaged by ASR after 50 years in service. *Engineering Structures*, 177, 133–146. https://doi.org/10.1016/j.engstruct.2018.09.043

Haha, M. Ben, Gallucci, E., Guidoum, A., & Scrivener, K. L. (2007). Relation of expansion due to alkali silica reaction to the degree of reaction measured by SEM image analysis. *Cement and Concrete Research*, 37(8), 1206–1214. https://doi.org/10.1016/j.cemconres.2007.04.016

ISE. (1992). *Structural effects of alkali-aggregate reaction: technical guidance on the appraisal of existing structures*. The Institution of Structural Engineers (ISE).

Kawabata, Y., Martin, R.-P., Seignol, J.-F., & Toutlemonde, F. (2016). Modelling of evolution of transfer properties due to expansion of concrete induced by internal swelling reaction. *5th International Conference on Alkali Aggregate Reaction*.

Kawamura, M., & Kodera, T. (2005). Effects of externally supplied lithium on the suppression of ASR expansion in mortars. *Cement and Concrete Research*, 35(3), 494–498. https://doi.org/10.1016/j.cemconres.2004.04.032

Kobayashi, K., & Takagi, Y. (2020). Penetration of pressure-injected lithium nitrite in concrete and ASR mitigating effect. *Cement and Concrete Composites*, 114, 103709. https://doi.org/10.1016/j.cemconcomp.2020.103709

Larive, C. (1997). *Apports combinés de l'expérimentation et de la modélisation à la compréhension de l'alcali-réaction et de ses effets mécaniques* [PhD]. Ecole Nationale des Ponts et Chaussées.

LCPC. (2003). *Aide à la gestion des ouvrages atteints de réactions de gonflement interne. Techniques et méthodes des laboratoires des ponts et chaussées, Guide méthodologique* (pp. 1–66). Laboratoire central des ponts et chaussées.

Martin, R. P. (2010). *Analyse sur structures modèles des effets mécaniques de la réaction sulfatique interne du béton*. Laboratoire central de ponts et chausses (LCPC).

Martin, R. P., Renaud, J. C., & Toutlemonde, F. (2010). Experimental investigations concerning combined delayed ettringite formation and alkali aggregate reaction. *6th International Conference on Concrete under Severe Conditions CONSEC10*.

Martin, R.-P., Sanchez, L., Fournier, B., & Toutlemonde, F. (2017). Evaluation of different techniques for the diagnosis & prognosis of Internal Swelling Reaction (ISR) mechanisms in concrete. *Construction and Building Materials*, 156, 956–964. https://doi.org/10.1016/j.conbuildmat.2017.09.047

Merz, C., & Leemann, A. (2013). Assessment of the residual expansion potential of concrete from structures damaged by AAR. *Cement and Concrete Research*, 52, 182–189. https://doi.org/10.1016/j.cemconres.2013.07.001

Moës, N., Cloirec, M., Cartraud, P., & Remacle, J.-F. (2003). A computational approach to handle complex microstructure geometries. *Computer Methods in Applied Mechanics and Engineering*, 192(28–30), 3163–3177. https://doi.org/10.1016/S0045-7825(03)00346-3

Multon, S., Barin, F.-X., Godart, B., & Toutlemonde, F. (2008). Estimation of the residual expansion of concrete affected by Alkali Silica reaction. *Journal of Materials in Civil Engineering, 20*(1), 54–62. https://doi.org/10.1061/(ASCE) 0899-1561(2008)20:1(54)

Multon, S., Sellier, A., & Cyr, M. (2009). Chemo-mechanical modeling for prediction of alkali silica reaction (ASR) expansion. *Cement and Concrete Research, 39*(6), 490–500. https://doi.org/10.1016/j.cemconres.2009.03.007

Nguyen, T. N., Sanchez, L. F. M., Li, J., Fournier, B., & Sirivivatnanon, V. (2022). Correlating alkali-silica reaction (ASR) induced expansion from short-term laboratory testings to long-term field performance: A semi-empirical model. *Cement and Concrete Composites, 134*, 104817. https://doi.org/10.1016/j.cemconcomp.2022.104817

Plusquellec, G., Geiker, M. R., Lindgård, J., Duchesne, J., Fournier, B., & De Weerdt, K. (2017). Determination of the pH and the free alkali metal content in the pore solution of concrete: Review and experimental comparison. *Cement and Concrete Research, 96*, 13–26. https://doi.org/10.1016/j.cemconres.2017.03.002

Ramlochan, T., Zacarias, P., Thomas, M. D. A., & Hooton, R. D. (2003). The effect of pozzolans and slag on the expansion of mortars cured at elevated temperature: Part I: Expansive behaviour. *Cement and Concrete Research, 33*(6), 807–814. https://doi.org/10.1016/S0008-8846(02)01066-9

Sanchez, L. F. M., Multon, S., Sellier, A., Cyr, M., Fournier, B., & Jolin, M. (2014). Comparative study of a chemo-mechanical modeling for alkali silica reaction (ASR) with experimental evidences. *Construction and Building Materials, 72*, 301–315. https://doi.org/10.1016/j.conbuildmat.2014.09.007

Saouma, E. V. (2021). Diagnosis & prognosis of AAR-affected structures state-of-the-art report of the RILEM technical committee 259-ISR. In *RILEM State-of-the-Art Reports*. Springer.

Sellier, A., Bourdarot, E., Multon, S., Cyr, M., & Grimal, E. (2009). Combination of structural monitoring and laboratory tests for assessment of Alkali-aggregate reaction swelling: Application to gate structure dam. *Combination of Structural Monitoring and Laboratory Tests for Assessment of Alkali-Aggregate Reaction Swelling: Application to Gate Structure Dam*, 281–290.

Souza, L. M. S., Polder, R. B., & Çopuroğlu, O. (2017). Lithium migration in a two-chamber set-up as treatment against expansion due to alkali-silica reaction. *Construction and Building Materials, 134*, 324–335. https://doi.org/10.1016/j.conbuildmat.2016.12.052

Stokes, D., Wang, H., & Diamond, S. (1997). A lithium-based admixture for ASR control that does not increase the pore solution pH. *Proceedings of the Fifth CANMET/ACI Inter-national Conference on Superplasticizers and Other Chemical Admixtures in Concrete: ACI SP-173* (pp. 855–867). Skokie, IL, American Concrete Institute.

Swamy, R. N. (1991). *The Alkali-Silica Reaction in Concrete* (R. N. Swamy, Ed.). CRC Press. https://doi.org/10.4324/9780203036631

Swamy, R. N. (1997). Assessment and rehabilitation of AAR-affected structures. *Cement and Concrete Composites, 19*(5–6), 427–440. https://doi.org/10.1016/S0958-9465(97)00035-8

Villemure, F. A., Fiset, M., Bastien, J., Mitchell, D., & Fournier, B. (2019). Behavior of epoxy bonded bars in concrete affected by alkali-silica reaction. *ACI Material Journal, 116*(6).

Zubaida, N. (2020). *Evaluation of the potential of residual expansion of concrete affected by alkali aggregate reaction* [Master]. University of Ottawa.

Chapter 9

Case study

Condition assessment of the Robert-Bourassa/ Charest (RBC) overpass

9.1 INTRODUCTION

The Robert-Bourassa/Charest (RBC) overpass, located in Quebec City, Canada, was a highway bridge structure constructed in 1966 using an alkali-silica reactive coarse limestone aggregate. The overpass consisted of a deck resting on Y-shaped reinforced concrete columns, which in turn were supported by massive concrete foundation blocks (Figure 9.1) (Fournier et al. 2015). While there is no specific information available regarding the concrete mix designs, technical reports indicate that the 28-day design strengths were 24 MPa for the foundation blocks and 28 MPa for the columns and decks (ICAAR Visit Report 2000). Over the course of the past three decades, the structural members of the RBC overpass have exhibited various signs of deterioration. These signs include extensive steel corrosion and concrete delamination and spalling at the deck level, along with map cracking, scaling, disaggregation and pop-outs affecting the foundation blocks due to alkali-silica reaction (ASR) and freeze and thaw (FT) cycles. Additionally, steel corrosion and concrete spalling have been observed on the columns and foundation blocks exposed to saltwater spray from traffic on the Robert-Bourassa highway (Bérubé et al. 2005). To address the concerns regarding the long-term performance of the Y-shaped columns of the RBC overpass, a series of rehabilitation techniques were implemented in the year 2000; these techniques aimed to mitigate further expansion and damage caused by ASR. The following is a detailed description of the selected products used for this purpose (Sanchez et al. 2020):

- *Waterproof polymer-modified sealant (WPMS)*: This hand-spray-applied material was used to prevent water leakage and moisture uptake into the structural member.
- *Cementitious coating*: A protective coating applied to the concrete surface to prevent moisture uptake and improve aesthetics.
- *Silane/siloxane sealants*: These water-repellent materials do not create a barrier but are effective in minimizing moisture uptake from the environment.

DOI: 10.1201/9781003188155-9

Figure 9.1 RBC overpass after nearly 50 years in service (Sanchez et al. 2020).

- *Glass fibre reinforced polymer (GFRP) sheet*: High-performance carbon sheets utilized to reinforce the distressed concrete members. They also serve to prevent water leakage and moisture uptake.
- *Epoxy coating*: A protective polymeric coating applied to concrete members to prevent moisture uptake.
- *Copolymer-modified coating*: This high-build, copolymer-modified coating is specially formulated to prevent water leakage and moisture uptake.

Through a series of site inspection surveys, field monitoring and laboratory tests conducted on the RBC overpass, a comprehensive assessment of the structure's condition was performed. The results of these investigations served as the basis for the development of a guide for the diagnosis, prognosis and management of ASR-affected structures proposed by Bérubé et al. and presented in Chapter 8 (Bérubé et al. 2005). The guide included a wide range of laboratory tests, such as the stiffness damage test (SDT), Damage Rating Index (DRI), residual expansion and water-soluble alkalis, among others. While the guide showed promise, it was found that certain test procedures, particularly the SDT and DRI, were found not to be fully diagnostic, limiting the efficiency of the proposed protocol. Additionally, the visual inspection analyses comparing the performance of the different rehabilitation products used on the columns did not yield conclusive results. Therefore, questions were raised regarding the efficiency of these products in mitigating further expansion and deterioration caused by ASR and FT cycles (Sanchez et al. 2020).

Before the demolition of the RBC overpass in 2010/2011, several cores were extracted from the various structural members, including the foundation blocks (FB), treated and untreated columns (C), and the bridge deck (BD). These cores allowed for a comprehensive multi-level assessment of the overall condition of the structure, following the protocols outlined in (Sanchez et al. 2014, 2015, 2016a, 2016b, 2017, 2018) and described in Chapter 7. In addition to the cores' extraction, a thorough visual inspection was conducted on all columns, both treated and untreated, prior to the extraction of the cores.

9.2 CONDITION ASSESSMENT OF RBC OVERPASS

9.2.1 Definition of exposure conditions (microclimate)

It should be noted that the RBC overpass consisted of two parallel bridge structures: the South and North Bridges. While the overall macroclimate conditions were considered similar for all members of the overpass, variations in microclimate conditions were observed. Considering this, the exposure conditions for the different members were defined as follows:

- *FB*: Distinct locations of the FB were categorized as either exposed (E) or non-exposed (NE). The E condition referred to locations directly exposed to weather elements such as wind, rain and splash zones. On the other hand, the NE condition denoted locations protected by the BD.
- *Columns*: The first column at the edge of the deck was designated as a highly exposed (HE) condition. The second and third columns towards the centre of the BD were classified as moderately exposed (ME), while the remaining columns were considered NE. Refer to Figure 9.2 for further details.

No specific exposure conditions were assigned to the BD itself since it was completely covered by an asphalt concrete layer.

9.2.2 Visual inspection and crack measurements

A visual inspection was conducted on all Y-shaped columns of the RBC overpass, including treated and untreated columns in both E and NE conditions. The objective of this inspection was twofold: (a) to compare the

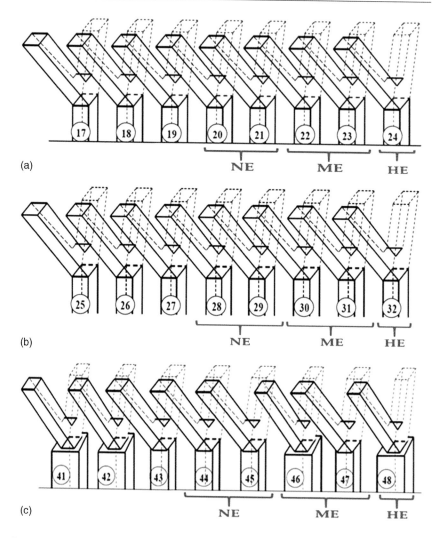

Figure 9.2 Identification of exposure conditions for columns: (a) north-west group, (b) south-west group and (c) south-east group (Sanchez et al. 2020).

different treatments applied to the columns and (b) to assess any potential disparity between surface damage (evaluated through visual inspection) and internal damage (assessed through microscopic and/or mechanical tests on extracted cores). Due to significant deterioration, a detailed visual inspection was deemed unnecessary for the other members (i.e., BD and FB), and coring was directly proceeded with.

During the visual inspection of RBC columns, a qualitative and semi-quantitative evaluation of the observed deterioration degree was performed.

Each column received a qualitative rating based on its overall visual condition using the following scale (Sanchez et al. 2020):

- 0: Undamaged
- 1: Very minor signs of damage
- 2: Noticeable signs of damage
- 3: Moderate damage
- 4: High damage
- 5: Very high damage

Furthermore, semi-quantitative measurements were conducted to assess the importance of cracks, specifically crack openings, in each column. Four very fine cracks and four wide-opened cracks were selected from each column arm to capture the "damage range" for each column. Table 9.1 provides a summary of the main results from the visual inspection of the RBC columns. Figure 9.3 depicts the average maximum crack opening values (average of the four wide-opened cracks) measured on the columns, while the complete range of crack openings is presented in Table 9.1. Figure 9.4 showcases the general condition of selected evaluated columns, such as columns 17, 25, 28 and 29.

9.2.2.1 Untreated columns

The evaluation of untreated columns on the RBC overpass revealed notable findings. HE columns from both the South and North structures, namely Y25, Y32, Y41, Y48 and Y17, displayed visual distress ratings ranging from high to very high, with values between 4 and 5, as shown in Table 9.1. Similarly, ME columns, including Y18, Y26 and Y42, exhibited significant visual distress ratings ranging from 3 to 4. Moreover, both HE and ME columns showed elevated average maximum crack openings, some reaching 1.5 mm–2.0 mm. In contrast, the untreated and NE columns, specifically Y19, Y20 and Y21, exhibited much less visual distress when compared to their ME and HE counterparts. The protection provided by their location shielded them from the detrimental effects of the environment, resulting in better visual conditions.

9.2.2.2 Treated columns

The visual assessment of the RBC overpass columns revealed interesting findings. The HE-Y25 column, treated with a waterproof sealant, exhibited a level of visual deterioration comparable to the untreated Y17 column. Both columns received a distress rating of 4 and displayed an average maximum crack opening of 1.8 mm. However, Y17 appeared wetter in appearance when compared to Y25, a characteristic previously reported by Bérubé et al. (Bérubé et al. 2002) after applying waterproof sealants,

Table 9.1 Visual inspection results from distinct RBC columns (Sanchez et al. 2020)

Group	Column	Treatment	Exposure class	Qualitative damage rating[a]	Crack opening (mm)[b]
South-West	Y25	Waterproof sealant	HE	4	0.60–3.0
	Y26	–	ME	3–4	1.00–4.0
	Y27	–	ME	3–4	0.60–1.0
	Y28	Waterproof sealant	NE	2	0.20–1.3
	Y29	Epoxy coating	NE	2	0.60–1.5
	Y30	Waterproof sealant + copolymer-modified coating	ME	1	0.10–0.6
	Y32	–	HE	5	0.60–1.0
South-East	Y41	–	HE	5	0.40–2.0
	Y42	–	ME	4	0.50–5.0
	Y43	–	ME	3	0.25–1.0
	Y46	–	ME	3	0.33–3.0
	Y48	–	HE	4–5	0.60–2.0
North-West	Y17	–	HE	4	0.40–2.0
	Y18	–	ME	4	1.0
	Y19	–	ME	2	0.80–2.0
	Y20	–	NE	2	0.40–0.8
	Y21	–	NE	2	0.40–0.6
	Y22	Waterproof sealant + Cementitious coating	ME	1–2	0.00–0.5
	Y23	Waterproof sealant + Silane/siloxane sealants	ME	1	0.00–0.1
	Y24	GFRP sheet	HE	0–1	0.00

[a] Definitions according to 4.1.
[b] Cracks range observed in the distinct columns.

specifically silane-based products, to ASR-affected members. In contrast, NE columns – namely, Y28 and Y29 – treated with a waterproof sealant and epoxy coating, respectively, showed some degree of deterioration but to a lesser extent than the HE-Y25 column. They received a qualitative rating of 2 and exhibited average maximum crack openings of 0.5 mm and 0.9 mm,

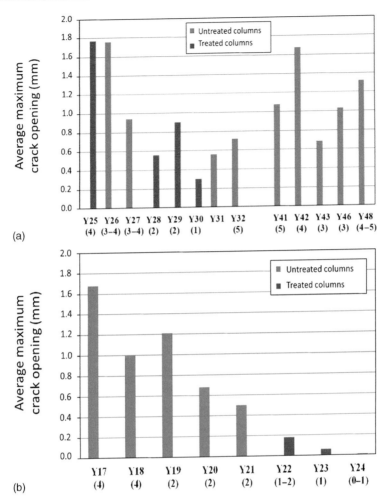

Figure 9.3 Maximum crack opening of distinct treated and untreated RBC columns: (a) south-west (left) and south-east (right) groups, (b) north-west group (Sanchez et al. 2020).

respectively. The observed damage on these NE columns closely resembled that of the untreated columns (i.e., Y19 and Y21). Moving on to the ME columns, including Y30, Y22 and Y23, they were initially treated with various products: waterproof sealant, copolymer-modified coating, cementitious coating and silane/siloxane-based products, respectively. All these columns displayed a low to very low qualitative visual damage level. Cracks with very small openings, measuring 0.3 mm, 0.2 mm and 0.1 mm, respectively, were observed on each of them. Lastly, the HE-Y32 column, which underwent GFRP wrapping, showed no noticeable surface cracking.

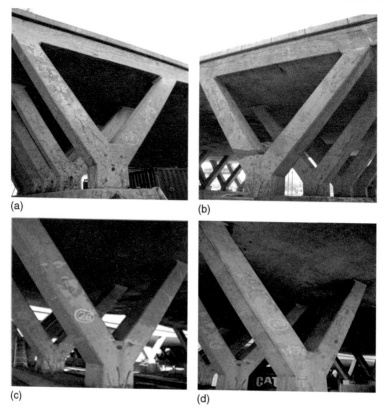

Figure 9.4 Visual aspect of RBC columns: (a) HE untreated Y17 column, (b) HE waterproof treated Y25 column, (c) NE waterproof treated Y28 column and (d) NE epoxy coating Y29 column (Sanchez et al. 2020).

9.2.3 Coring

Core samples were carefully extracted from the various members of the RBC overpass, including the FB, columns (C), and BD of both the South and North structures. In the case of the BD, the coring process involved the extraction of three large samples with a diameter of 800 mm (Figure 9.5a). These large cores were then stored at a controlled temperature of 23 ± 2°C for 25 weeks. Subsequently, a second coring step was conducted, resulting in smaller cores with a diameter of 100 mm and a length of 200 mm. These smaller cores were obtained from three directions: vertical (V), transversal (T) and longitudinal (L) within each large sample (Figure 9.5b). From each core, three specimens measuring 100 by 200 mm were obtained per direction, as shown in Figure 9.6a (dashed lines). It should be noted that the first inch of each core from both ends was discarded due to surface flaws and defects resulting from the coring process.

Figure 9.5 (a–f) Coring from BD, columns and FB (Sanchez et al. 2020).

Furthermore, ten distinct columns from the South and North structures, numbered 17, 22, 23, 24, 25, 28, 29, 30, 32 and 42, were evaluated. These columns underwent different treatments and were exposed to varying conditions (Table 9.1). After the demolition of the RBC overpass, some columns were saw-cut and stored for approximately two weeks prior to coring (Figure 9.5c and d). Eight core specimens were extracted from each column, and the samples' cover was subsequently removed. Therefore, a single specimen per core was obtained, as depicted in Figure 9.6b; only the "internal" specimens located within the reinforcement cage of the columns were evaluated.

In the case of the FB, coring was conducted at exposed and NE locations. Eighteen samples were extracted per location in a semi-circular manner, as shown in Figure 9.5e and f. Similar to the BD, these samples were saw-cut, discarding the first inch at each side (Figure 9.6c). Consequently, two specimens measuring 100 by 200 mm were obtained per core per location.

Upon extraction, all 100 mm diameter cores were carefully wrapped in plastic film to prevent moisture loss and stored at a temperature of 12°C. This storage condition was implemented to halt any further deterioration of the cores, as recommended by (Sanchez et al. 2017). Prior to conducting mechanical testing, the specimens were saw-cut to a length of 200 mm and then placed in a moist curing room at a temperature of 23 ± 2°C and 100% relative humidity for 48 hours. The ends of all specimens were ground using a mechanical grinder to ensure uniformity. The samples for petrography via the DRI were axially cut into two halves, and one of the resulting flat surfaces was polished. A portable hand-polishing device equipped with diamond-impregnated rubber disks was utilized to achieve the desired surface quality for microscopic assessment. The disks used ranged in coarseness from no. 50 (coarse) to 1,500, 3,000 (very fine). This approach proved effective in obtaining surfaces suitable for detailed microscopic examination.

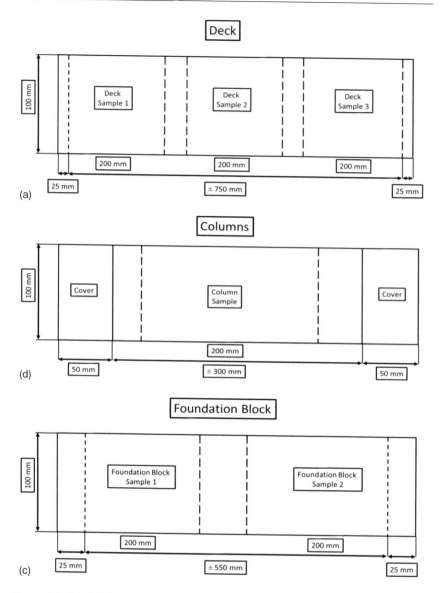

Figure 9.6 (a–c) Schematics of cores preparation (cutting and grinding) for testing (Sanchez et al. 2020).

Eight 100 mm by 200 mm samples per member per condition were selected for assessment of RBC members, as described in Table 9.2.

9.2.4 Laboratory test methods

Following coring, the multi-level assessment as per Sanchez et al. (2017, 2018) was conducted; it is important to mention that conventional petrography to

Table 9.2 Number of cores extracted from the different RBC members (Sanchez et al. 2020)

Concrete members	Condition			
	E (surface)	E (core)	NE (surface)	NE (core)
FB	8	8	8	8

	Longitudinal		Vertical		Transversal		
	T^a	B^b	T^a	B^b	T^a	B^b	BD
8	8	8	8	8	8		

	Exposure Class	Treatment	Samples
Column 17	HE	UN	8
Column 22	ME	T; Waterproof sealant + Cementitious coating	8
Column 23	ME	T; Waterproof sealant + Silane/siloxane sealants	8
Column 24	HE	T; GFRP sheet	8
Column 25	HE	T; Waterproof sealant	8
Column 28	NE	T; Waterproof sealant	8
Column 29	NE	T; Epoxy coating	8
Column 30	ME	T; Waterproof sealant + copolymer-modified coating	8
Column 32	HE	UN	8
Column 42	NE	UN	8

[a] T: Top of the BD;
[b] B: Bottom of the BD.

detect the cause(s) of deterioration was not performed because ASR and FT had already been detected in RBC members in previous evaluations as per Bérubé et al. 2005. Therefore, it has been decided to proceed directly with the multi-level protocol according to the tests presented hereafter:

- *SDT*: The SDT was employed to quantify the damage degree of the distinct cores extracted from the various RBC members. The SDT involves subjecting concrete specimens to five compressive cycles, using 40% of the compressive strength of a companion sound concrete with similar features and maturity. Chapter 6 provides a complete description of the test procedure. Concrete specimens from each set of cores, as outlined in Table 9.2, underwent five loading-unloading cycles at a controlled loading rate of 0.10 MPa/s. Each set consisted of three concrete cores obtained from the same structural member, with similar environmental conditions, depth from the surface, direction and treatment type.

- *DRI*: The DRI was also used to assess the damage degree of the distinct cores extracted from the various RBC members. The DRI involves microscopically analysing polished concrete sections under a stereomicroscope with a magnification of 15x–16x. Petrographic damage features observed are then counted within 1 cm^2 grids drawn on the polished concrete sections. Weighting factors are multiplied by the feature counts to balance their relative importance towards the overall deterioration of the material. Finally, the DRI number is computed; the higher the DRI number, the higher the deterioration of the concrete. Ideally, a surface of at least 200 cm^2 should be considered per concrete member, yet for comparative purposes, the final DRI number is normalized to 100 cm^2. Chapter 5 provides a comprehensive description of the DRI procedure. The DRI analysis was conducted on two companion cores (i.e., 100 cm^2 each – total of 200 cm^2) from each set of cores extracted from the same structural member, with similar environmental conditions, depth from the surface, direction and treatment type, as specified in Table 9.2.
- *Compressive strength tests*: Although the compressive strength of concrete is less affected by internal swelling reaction (ISR), particularly ASR, when compared to properties such as tensile strength and modulus of elasticity (Bérubé et al. 2002; ISE 1992; Nixon & Bollinghaus 1985; Smaoui et al. 2004; Wood & Johnson 1993), compressive strength tests were performed on the cores extracted from the different RBC members. The purpose of these tests was twofold: (a) first, to determine the ultimate strength capacity of non-damaged or less-damaged samples from the distinct members, which served as the basis for the 40% value used in the SDT, and (b) second, to evaluate any structural implications of ASR-induced damage over time on the RBC members. Therefore, two initial concrete cores were extracted from undamaged or less-damaged members/locations from the columns, BD and FB and compressive strength tests were conducted. Furthermore, compressive strength tests were also performed on all concrete cores obtained from the distinct RBC-affected members after stiffness damage testing. The latter has been shown to be valid by Sanchez et al. (2014).

9.3 RESULTS

9.3.1 Stiffness damage test (SDT)

9.3.1.1 Untreated members

The results of the SDT, including the Stiffness Damage Index (SDI), Plastic Deformation Index (PDI), modulus of elasticity (E) reductions in percentage and E values in GPa, are depicted in Figure 9.7. Upon analysing the data,

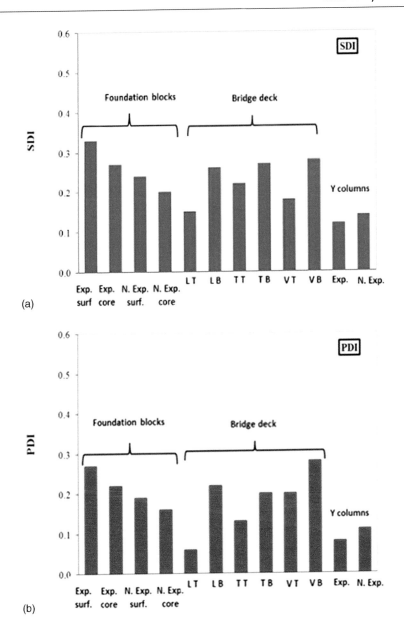

Figure 9.7 (a) SDI, (b) PDI.

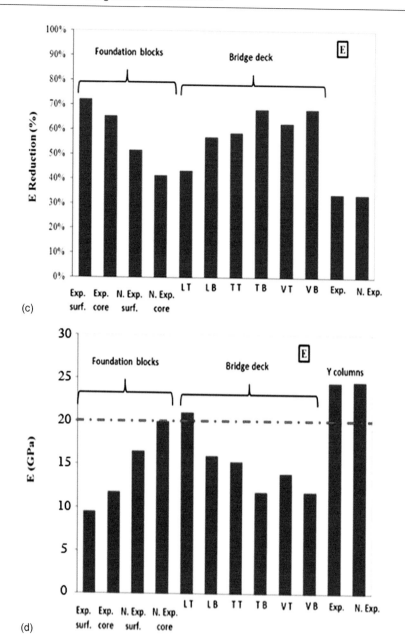

(c)

(d)

Figure 9.7 (Continued) (c–d) E results (E reduction, % – left; E values, GPa – right) from RBC untreated members (Sanchez et al. 2020).

it is evident that the FB specimens exhibited higher SDI and PDI values, as well as greater E reductions and lower E values when compared to the BD and C cores. The results also varied depending on the specimen's condition, such as the type of environment and depth from the surface. Notably, the exposed FB cores demonstrated higher SDI and PDI values when compared to the NE specimens. This observation aligns with the fact that the presence of high relative humidity or moisture is a prerequisite for ISR development. Additionally, surface specimens displayed more significant damage than internal samples. For the FB samples, the SDI values ranged from 0.20 to 0.35, while the PDI results varied from 0.16 to 0.25.

Different levels of damage were observed in the BD cores, depending on their orientation. Cores oriented vertically (V) and transversally (T) with respect to traffic exhibited higher damage, as indicated by higher SDI, PDI, greater E reductions and lower E values. In contrast, longitudinal (L) specimens showed lower levels of deterioration. Additionally, cores taken from the top of the deck (LT, TT, VT) displayed less damage when compared to those taken from the bottom (LB, TB, VB). In the case of the columns (C), the damage degree was lower than that of the FB and BD cores, with SDI ranging from 0.12 to 0.15 and PDI from 0.08 to 0.12. This can be attributed to the amount of restraint due to reinforcement, as well as the compression service loading experienced by these members.

Overall, the FB cores exhibited a very high degree of damage, as classified by Sanchez et al. (2017, 2018)), while the BD cores displayed moderate damage and C cores a marginal deterioration. It is worth noting that several FB and BD cores displayed modulus of elasticity values below 20 GPa, which is significantly low compared to expected values (34–37 GPa, depending on the member) for sound concrete with similar strength and aggregate type (Sanchez et al. 2017).

9.3.1.2 Treated and untreated columns

The SDT results obtained from the treated and untreated columns of the RBC overpass are presented in Figure 9.8. Upon analysis, it can be observed that the North-West specimens (i.e., 14, 22, 23 and 24) exhibited higher SDI and PDI values when compared to the South-West (i.e., 25, 28, 29, 30 and 32) and South-East (i.e., 42) cores. Surprisingly, the untreated (UN) and HE columns (i.e., 17, 32 and 42) did not necessarily display higher levels of damage, despite the expectation that the presence of high relative humidity (or moisture) would contribute to ASR-induced development. The SDI values for the HE untreated columns were 0.15, 0.14 and 0.12, while the PDI results were 0.12, 0.11 and 0.08, respectively.

Among all the members, columns 22, 23 and 24 exhibited the highest SDI and PDI values, with values of 0.16, 0.18 and 0.17 for SDI and 0.10, 0.15, and 0.10 for PDI, respectively. These columns were treated (T) and classified as ME/HE, according to Table 9.2. Columns 25 and 30 also showed

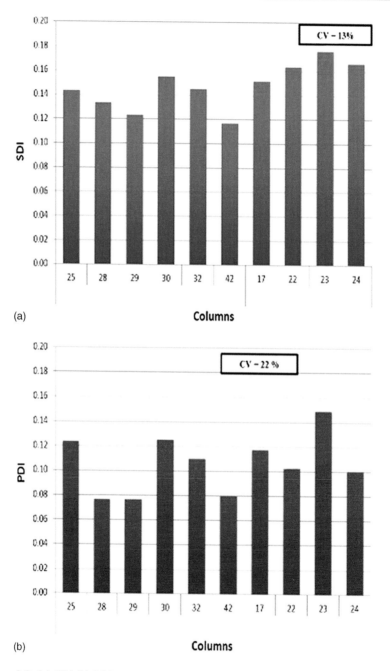

(a)

(b)

Figure 9.8 (a) SDI, b) PDI.

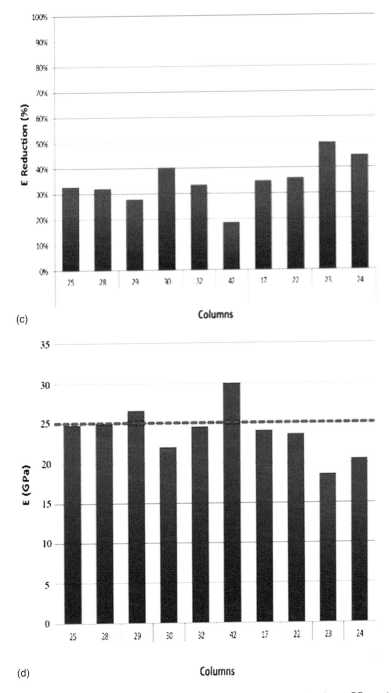

(c)

(d)

Figure 9.8 (Continued) (c-d) E results (E reduction, % – left; E values, GPa – right) from RBC treated and untreated columns (Sanchez et al. 2020).

relatively high SDI and PDI values, with values of 0.14, 0.15 and 0.12 for SDI and 0.13 for PDI. These columns were treated and classified as HE and ME, respectively. Finally, members 28 and 29 displayed the lowest SDI and PDI results (except for the SDI value of column 42), with values of 0.13 and 0.12 (SDI) and 0.07 (PDI), respectively. These columns were treated and categorized as NE.

In general, the damage levels observed in most columns were classified as marginal to moderate, according to Sanchez et al. (2017, 2018)). Additionally, the damage levels in the untreated (UN) or treated (T) columns were much lower compared to the damage observed in the FB and BD cores. The modulus of elasticity values obtained exhibited a strong correlation with the SDI and PDI values, indicating that higher SDI and PDI values were associated with lower modulus of elasticity values. The modulus of elasticity (E) values ranged from 30 GPa (in the case of column 42, which was untreated and HE) to 17 GPa (in the case of column 23, which was treated and ME). This represents a stiffness decrease ranging from 20% to 50%, with an average decrease of approximately 30%. From a structural engineering perspective, such a decrease in stiffness can be considered significant.

9.3.2 Damage rating index (DRI)

9.3.2.1 Untreated members

Figure 9.9 depicts the microscopic damage features and DRI numbers observed on cores extracted from the distinct untreated members of RBC (i.e., FB, BD and C). Upon examination of the plots, it becomes evident that all specimens exhibited closed cracks within the aggregate particles (CCA).

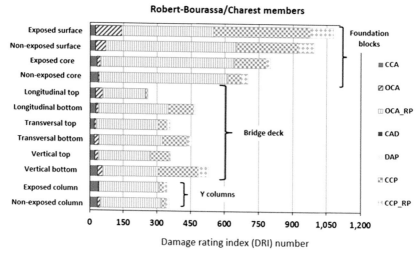

Figure 9.9 DRI results from RBC untreated members (Sanchez et al. 2020).

These CCA cracks are likely a result of aggregate production processes (e.g., crushing, sieving) and weathering and may not be directly associated with ISR-induced deterioration. However, the presence of open cracks with and without gel within the aggregate particles (OCA and OCA_{RP}, respectively) indicates ISR deterioration, particularly caused by ASR. Similarly, cracks in the cement paste with and without gel (CCP and CCP_{RP}) are also attributed to ASR. The DRI numbers obtained align with the SDT results, as higher values are observed in the FB specimens, followed by BD and C cores. Once again, the extent of damage varied depending on the condition of the cores. Exposed FB cores exhibited more damage when compared to NE specimens, while surface specimens displayed greater damage than internal samples.

In the case of BD cores, the damage levels varied according to the orientation of the cores. Vertical (V) cores demonstrated greater damage when compared to longitudinal (L) and transverse (T) specimens. These findings slightly differ from the mechanical analyses, where T specimens showed higher damage than L cores. Additionally, specimens from the top of the BD displayed lower damage when compared to specimens from the bottom. The C cores exhibited lower damage degrees than FB and BD cores, which can be attributed to the amount of restraint due to reinforcement present in these members.

9.3.2.2 Treated and untreated columns

Figure 9.10 provides a visual representation of the microscopic damage features and DRI numbers observed on cores extracted from the distinct treated and untreated columns of RBC. It is important to note that the analysis focused on samples located within the reinforcement cage, disregarding the columns' cover, as shown in Figure 9.3b. Upon examination, it is evident that all specimens exhibited CCA, which, as previously mentioned, is not necessarily indicative of ISR-induced damage. However, the presence of opened cracks with and without gel within the aggregate particles (OCA and OCA_{RP}, respectively) and the cement paste (CCP and CCP_{RP}) are indeed features associated with ISR (i.e., ASR more specifically) and were observed in all the specimens extracted from various columns. The DRI numbers obtained for both the treated and untreated columns ranged from 200 to 400, indicating a marginal to moderate level of damage in accordance with (Sanchez et al. 2017, 2018), which aligns with the mechanical findings obtained from the SDT analysis.

The DRI numbers obtained from the cores extracted from the North-West group exhibited a range of 200 to 300. Minimal variations were observed between individual cores within this group. On the other hand, the South-West/East columns displayed slightly higher average DRI values, ranging from 200 to 400. Notably, the highly and ME columns (e.g., 25 and 30) generally exhibited greater damage when compared to the NE columns

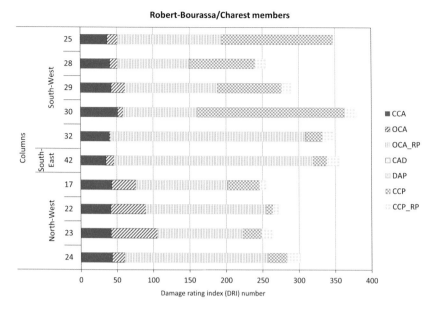

Figure 9.10 DRI results from the distinct RBC treated and untreated columns (Sanchez et al. 2020).

(e.g., 28 and 29). Furthermore, a notable presence of cracks within the cement paste was observed, particularly in comparison to the North-West group, suggesting a more pronounced manifestation of ISR-induced development in this group of columns. It is plausible that the North-West columns experienced a combination of mechanisms such as ASR and FT, resulting in cracks within the aggregates and cement paste. Overall, the values obtained from the different untreated (UN) or treated (T) columns were lower than the damage levels observed in the FB and BD cores.

9.3.3 Compressive strength results

Figure 9.11 illustrates the obtained compressive strength results of the cores extracted from the various members of the RBC structure. The compressive strengths ranged from 20 to 33 MPa for the BD, 20 to 28 MPa for the FB and 33 to 42 MPa for the different columns. Additionally, it was observed that the compressive strengths of the BD cores varied depending on their orientation. Specifically, the vertical cores exhibited slightly lower compressive strength values in comparison to the transverse and longitudinal samples.

Half of the evaluated cores from the FB, specifically four out of eight cores (50%), exhibited compressive strength values lower than the 24 MPa design values. Similarly, most of the samples from the BD, with 10 out of 17 samples (58%), displayed compressive strength values below the 28 MPa design value. On the other hand, all the cores from the C demonstrated compressive strengths exceeding the 28 MPa design value.

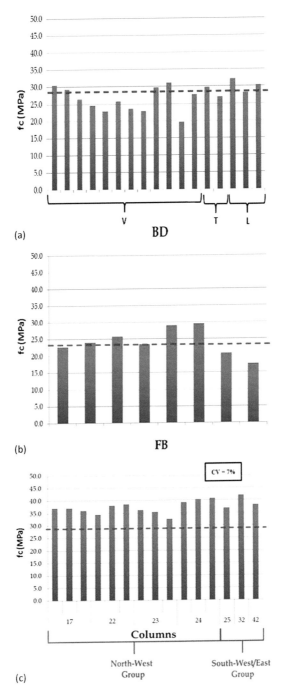

Figure 9.11 Compressive strength results for the distinct RBC members: (a) Bridge deck, (b) Foundation block and (c) Columns (Sanchez et al. 2020).

9.4 DISCUSSION

9.4.1 Multi-level assessment application

A comprehensive approach to appraise ISR-induced damage in concrete was developed by (Sanchez et al. 2017, 2018), particularly for alkali-aggregate reaction (AAR), DEF and FT deterioration, single or combined. Through an extensive analysis of various concrete strengths (25, 35 and 45 MPa), aggregate types (fine vs coarse) and natures (distinct lithotypes), a chart of data envelopes was formulated (Table 9.3) with a 95% confidence level. It is important to note that the chart shows some overlap of data, both microscopic and mechanical, which can be attributed to the inherent heterogeneity of concrete and its constituents, particularly the type and nature of reactive aggregates (Figure 9.12a – envelopes for 35 MPa concrete illustrating 12 distinct ASR reactive aggregates). To address this variability, Sanchez et al. (2017, 2018) suggested two approaches for utilizing the chart. Firstly, practitioners may select data points from similar reactive aggregates present in the concrete under examination (e.g., reactive limestone aggregate in this case). Secondly, if the reactive aggregate is unknown, averaging the envelope values can be considered. It should be noted that Table 9.3 was established based on laboratory test samples under free-expansion conditions, while structures and structural members experience various stress states and confinement conditions. Nonetheless, this chart serves as a valuable reference, providing an assessment of the "worst-case scenario" for ISR-affected concrete.

The strength of employing the multi-level approach lies in its correlation with ISR-induced expansion, offering engineers responsible for ageing infrastructure insights into the current state of the chemical reaction and its potential for further development. Such information is crucial for selecting appropriate rehabilitation techniques for damaged structural components. Additionally, it is important to highlight that Table 9.3 specifically presents values obtained from AAR-affected concrete. Data from

Table 9.3 Multi-level assessment results of AAR-affected concrete (Sanchez et al. 2017)

Classification of ASR damage degree (%)	Reference expansion level (%)	Assessment of ASR				
		Stiffness loss (%)	Compressive strength loss (%)	Tensile strength loss (%)	SDI	DRI
Negligible	0.00–0.03	-	-	-	0.06–0.16	100–155
Marginal	0.04 ± 0.01	5–37	(-)10–15	15–60	0.11–0.25	210–400
Moderate	0.11 ± 0.01	20–50	0–20	40–65	0.15–0.31	330–500
High	0.20 ± 0.01	35–60	13–25	45–80	0.19–0.32	500–765
Very high	0.30 ± 0.01	40–67	20–35		0.22–0.36	600–925

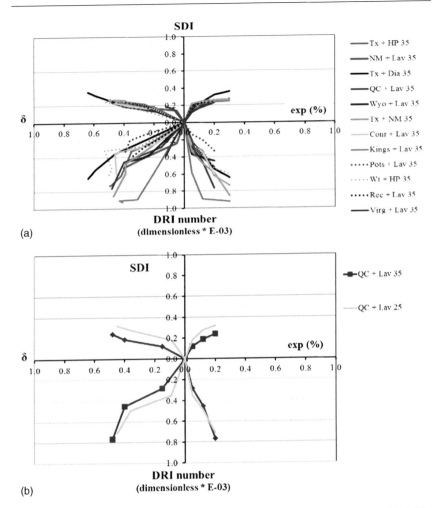

(a)

(b)

Figure 9.12 Multi-level assessment plots proposed by Sanchez et al. for (a) AAR-affected 35 MPa mixtures containing 12 distinct reactive aggregates and (b) AAR-affected 25 and 35 MPa mixtures incorporating a reactive limestone aggregate similar to the aggregate used in RBC members (Sanchez et al. 2020).

other ISR mechanisms, such as DEF and FT, either individually or in combination with AAR, have also been collected and documented in the literature, as discussed by Sanchez et al. (2018)).

9.4.1.1 Untreated members

Upon examining the microscopic and mechanical results obtained in this study (Figures 9.7–9.11) and comparing them with the data presented in

Table 9.3 and Figure 9.12b, which represent an AAR-affected concrete database utilizing a reactive limestone aggregate, it becomes evident that the FB, BD and C members exhibit varying degrees of damage (Sanchez et al. 2017, 2018, 2020). Specifically, FB is characterized by a very high level of damage, while BD and C display high and moderate levels of deterioration, respectively. These observations align with the "free-expansion" values, which are approximately 0.30%, 0.20% and 0.08% for the FB, BD and C, respectively.

Analysis of the DRI results reveals the significant presence of cracks within the aggregate particles, both with and without gel, indicating ASR as the primary cause of the ongoing damage process, as anticipated. Moreover, substantial cracks in the cement paste, partially filled with gel, are observed, particularly in the FB specimens. This suggests not only high levels of ASR development but also the likelihood of combined mechanisms such as freezing and thawing.

From an engineering standpoint, the compressive strength values obtained from the untreated members raise concerns. A notable portion of the FB and BD specimens exhibit compressive strength results lower than the expected design values, which brings doubts on the overpass's ability to withstand service loads as intended and in accordance with Canadian standards and safety protocols. Additionally, most of the untreated specimens display very low stiffness, as indicated by their modulus of elasticity results. This low stiffness may pose potential issues regarding the structure's serviceability, including deflection, displacement or even local deformation of members. Previous visual inspections have already identified significant deflections in certain bridge members, which can be partially attributed to the substantial decrease in the material's stiffness.

Furthermore, the values obtained from untreated members suggest that the unrestrained concrete expansion in some elements surpasses the yielding threshold of the steel reinforcement (0.20%), raising structural concerns. Additionally, as emphasized by Sanchez et al. (2017)), ASR cracks may diminish the aggregate interlock effects in damaged concrete due to their microscopic distress features, such as cracks splitting aggregate particles. Consequently, the overall shear capacity of affected structural members may be reduced. Therefore, the proper management of ASR-affected infrastructure necessitates in-depth evaluations of its potential structural implications. It is important to note that there is currently a lack of literature data and research on this subject matter.

9.4.1.2 Treated columns

The expansion levels observed in the distinct RBC columns, which underwent various rehabilitation procedures, ranged from 0.05% to 0.12%, as depicted in Figure 9.13 based on Table 9.3 and Figure 9.12b. These expansion values indicate a marginal to moderate degree of damage, on average. The North-West columns group exhibited higher levels of expansion compared to the

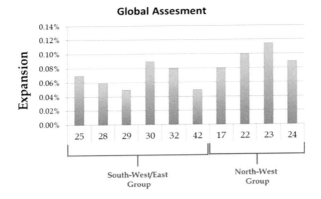

Figure 9.13 Potential expansion attained for the different RBC columns (Sanchez et al. 2020).

South-West/East group. The efficiency of the distinct treatment procedures could not be clearly determined in this study due to the low damage state of the columns and the lack of knowledge regarding the "0" damage values prior to treatment. However, it can be observed that, for the majority of the columns, a more exposed condition resulted in higher expansion levels. For instance, columns 22, 23 and 24 exhibited greater expansion when compared to columns 28 and 29. Furthermore, when comparing these results with the previously obtained data from untreated FB and BD members, it is evident that the deterioration found in the treated and untreated columns is significantly lower.

Although most of the treated and untreated columns in the North-West and South-West/East groups can be classified as having marginal to moderate damage according to Table 9.3, the specific damage features observed in each group of columns provide valuable insights into their distinct distress processes. This can be readily observed by analysing the DRI bar charts presented in Figures 9.9 and 9.10. The relatively low number of cracks in the cement paste (indicated by the checker an initial stage of ASR). In contrast, the higher occurrence of cement paste cracks in the South-West/East group indicates a greater degree of deterioration. However, upon closer examination of the charts for the South-West/East group, it becomes evident that although the cement paste cracks are more pronounced on average, the cracks in the aggregate particles are less prominent. This suggests the presence of another mechanism, such as FT, contributing to the overall deterioration process of the concrete in combination with ASR.

The evaluation of compressive strength results obtained from the treated columns reveals that most of these members exhibit compressive values higher than anticipated. This suggests that their structural capacity is less of a concern when compared to the FB and BD members. Nevertheless, the previous evaluations and discussions indicate that the bridge did not comply with the current Canadian standards and safety regulations at the time of its demolition in 2010/2011.

9.4.2 Visual vs multi-level assessment

Visual Inspection is typically employed as the initial non-destructive technique for inspecting concrete infrastructure. While it is often considered descriptive, qualitative and somewhat subjective, it plays a crucial role in decision-making and determining areas that require further assessment. In this study, both untreated and treated columns of RBC were subjected to visual inspection and multi-level appraisal, enabling a comparison between these two techniques. The outcomes obtained from the non-destructive visual inspection and the destructive microscopic and mechanical assessments are presented in Table 9.4.

Upon analysing the data presented in Table 9.4, it is observed that out of the ten columns assessed, only two consistent results were found between visual inspection and multi-level assessments (columns 28 and 29). Additionally, two types of inconsistencies were identified: visual inspection yielded higher damage levels than the multi-level assessment in four columns (25, 32, 42, 17), while in four other columns (30, 22, 23, 24), visual inspection indicated a lower damage degree when compared to the multi-level appraisal. These results suggest a weak correlation between the surface

Table 9.4 Comparison between visual and multi-level assessments (Sanchez et al. 2020)

		Visual inspection			Multi-level assessment		
Group	Column	Qualitative rating	Crack opening (mm)	Damage degree	Expansion (%)	Damage degree	Classification
South-West	Y25	4	0.60–3.0	High	0.07	Marginal	✗
	Y28	2	0.20–1.3	Marginal	0.06	Marginal	✓
	Y29	2	0.60–1.5	Marginal	0.05	Marginal	✓
	Y30	1	0.10–0.6	Undamaged	0.09	Moderate	✗
	Y32	5	0.60–1.0	Very high	0.08	Marginal/Moderate	✗
South-East	Y42	4	0.50–5.0	High	0.05	Marginal	✗
North-West	Y17	4	0.40–2.0	High	0.08	Moderate	✗
	Y22	1 to 2	0.00–0.5	Minor/Marginal	0.10	Moderate	✗
	Y23	1	0.00–0.1	Minor	0.11	Moderate	✗
	Y24	0 to 1	0	Undamaged/minor	0.09	Moderate	✗

condition and the internal damage degree of ISR-affected concrete. Therefore, the extent of surface cracking should not be solely relied upon to evaluate the inner condition and performance of ISR-affected concrete members. Nonetheless, this does not mean that surface cracking (and thus visual inspection) should not be considered whenever rehabilitation procedures are requested, especially when aiming to inhibit further durability-related issues such as FT and steel corrosion.

9.4.3 Potential structural implications on ISR-affected columns

9.4.3.1 Stress state of the stirrups

Three-metre sections were taken from two reinforced columns – one from an exposed site (E) and the other from a NE location to assess the potential structural implications of ISR on the affected columns, which posed significant concerns for the structural engineers overseeing the RBC project, as depicted in Figure 9.14. It is important to note that the E column had undergone FRP sheet wrapping in 2000 due to the severity of its deterioration, as illustrated in Figure 9.14b. In the laboratory, the concrete cover of the columns was removed at selected locations and the stirrups sanded and cleaned; then, strain gauges were glued on the stirrups to measure their stress relief upon cutting. A total of six stirrups were selected for testing at each column.

The stress relief in terms of micro strains ($\mu\varepsilon$) observed after cutting the instrumented stirrups from both columns is illustrated in Figure 9.15. It is important to note that the columns had been stored in the laboratory for several weeks prior to testing, during which no live loads were applied to the members. Therefore, it can be reasonably assumed that the strains resulting from stress relief of the stirrups are primarily attributed to ISR-induced expansion (ASR and FT in this case). The stress-state assessment of the stirrups' cutting was conducted in two ways. Initially, the stirrups were cut at their edges following the removal of the concrete cover. However, as the stirrups were often bent at the edges and sometimes tied to the columns' ends,

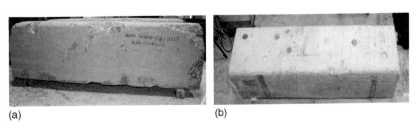

(a)   (b)

Figure 9.14 RBC columns: (a) NE column and (b) E column.

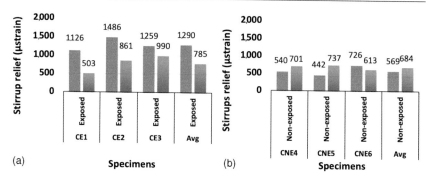

Figure 9.15 Stress relief measured on the two columns after stirrups cutting: (a) E column and (b) NE column. Blue bars represent the stirrups' edges, and red bars represent the centre of the cross-section.

analyses were also performed at the centre of the column's cross-section to determine if higher and potentially more reliable values could be obtained.

The strain values observed in the stirrups of the NE column were lower when compared to those of the E column, particularly when the analyses were conducted at the centre of the cross-section (indicated by the orange bars). When comparing data from the edge cuttings (blue bars) and centre cuttings (orange bars), higher values were found at the centre for the E column, while the edge values were relatively similar for both columns. This indicates that more accurate and reliable results are obtained from the centre cuttings of the stirrups. The average strain values obtained from the centre cuttings of the NE and exposed columns were 1290 and 570 µstrain, respectively. These results, especially for the E column, are considered quite high, particularly considering that the columns had been stored in the laboratory before testing, with no live loads applied that could contribute to the stress state of the stirrups at the time of cutting. Therefore, these values can be directly attributed to the development of ISR and could have been even higher if the cutting had been evaluated under actual service conditions. This emphasizes not only the well-known durability and serviceability concerns associated with ISR but also the potential structural implications caused by ISR-induced expansion and damage in concrete structures.

REFERENCES

Bérubé, M.-A., Chouinard, D., Pigeon, M., Frenette, J., Rivest, M., & Vézina, D. (2002). Effectiveness of sealers in counteracting alkali-silica reaction in highway median barriers exposed to wetting and drying, freezing and thawing, and deicing salt. *Canadian Journal of Civil Engineering*, 29(2), 329–337. https://doi.org/10.1139/l02-010

Bérubé, M.-A., Smaoui, N., Fournier, B., Bissonnette, B., & Durand, B. (2005). Evaluation of the expansion attained to date by concrete affected by alkali-silica reaction. Part III: Application to existing structures. *Canadian Journal of Civil Engineering*, *32*(3), 463–479. https://doi.org/10.1139/l04-104

Fournier, B., Sanchez, L., Beauchemin, S., & Au doctorat, candidat. (2015). *Outils d'investigation de la réactivité alcalis-granulats dans les infrastructures en béton Rapport Final par.*

ICAAR Visit Report. (2000). Report of the visit of structures affected by AAR in the Quebec City area. In *11th International conference on alkali-aggregate reaction*, Québec, Canada.

ISE. (1992). *Structural effects of alkali-aggregate reaction: technical guidance on the appraisal of existing structures*. The Institution of Structural Engineers (ISE).

Nixon, P. J., & Bollinghaus, R. (1985). The effect of alkali-aggregate reaction on the tensile strength of concrete. *Durable Construction Material*, *2*(3), 243–248.

Sanchez, L. F. M., Drimalas, T., Fournier, B., Mitchell, D., & Bastien, J. (2018). Comprehensive damage assessment in concrete affected by different internal swelling reaction (ISR) mechanisms. *Cement and Concrete Research*, *107*(February), 284–303. https://doi.org/10.1016/j.cemconres.2018.02.017

Sanchez, L. F. M., Fournier, B., Jolin, M., & Bastien, J. (2014). Evaluation of the stiffness damage test (SDT) as a tool for assessing damage in concrete due to ASR: Test loading and output responses for concretes incorporating fine or coarse reactive aggregates. *Cement and Concrete Research*, *56*, 213–229. https://doi.org/10.1016/j.cemconres.2013.11.003

Sanchez, L. F. M., Fournier, B., Jolin, M., Bastien, J., & Mitchell, D. (2016a). Practical use of the Stiffness Damage Test (SDT) for assessing damage in concrete infrastructure affected by alkali-silica reaction. *Construction and Building Materials*, *125*, 1178–1188. https://doi.org/10.1016/j.conbuildmat.2016.08.101

Sanchez, L. F. M., Fournier, B., Jolin, M., & Duchesne, J. (2015). Reliable quantification of AAR damage through assessment of the Damage Rating Index (DRI). *Cement and Concrete Research*, *67*, 74–92. https://doi.org/10.1016/j.cemconres.2014.08.002

Sanchez, L. F. M., Fournier, B., Jolin, M., Mitchell, D., & Bastien, J. (2017). Overall assessment of Alkali-Aggregate Reaction (AAR) in concretes presenting different strengths and incorporating a wide range of reactive aggregate types and natures. *Cement and Concrete Research*, *93*, 17–31. https://doi.org/10.1016/j.cemconres.2016.12.001

Sanchez, L. F. M., Fournier, B., Mitchell, D., & Bastien, J. (2020). Condition assessment of an ASR-affected overpass after nearly 50 years in service. *Construction and Building Materials*, *236*, 117554. https://doi.org/10.1016/j.conbuildmat.2019.117554

Sanchez, L., Fournier, B., Jolin, M., Bedoya, M. A. B., Bastien, J., & Duchesne, J. (2016b). Use of damage rating index to quantify alkali-silica reaction damage in concrete: Fine versus coarse aggregate. *ACI Materials Journal*, *113*(4). https://doi.org/10.14359/51688983

Smaoui, N., Bérubé, M. A., Fournier, B., & Bissonnette, B. (2004). Influence of specimen geometry, orientation of casting plane and mode of concrete consolidation on expansion due to ASR. *Cement, Concrete and Aggregates*, *26*(2), 58–70.

Wood, G. M., & Johnson, R. A. (1993). *The appraisal and maintenance of structures with alkali-silica reaction*. The Institution of Structural Engineers (ISE).

Chapter 10

Conclusions and future works

10.1 CONCLUSIONS, CURRENT CHALLENGES AND FUTURE OPPORTUNITIES

Internal swelling reactions (ISRs) are extremely complex damage mechanisms leading to induced expansion and deterioration of affected concrete. If, on the one hand, Chapters 1 and 2 demonstrate the complicated scientific nuances (yet to be fully understood) of the most common ISR mechanisms in concrete, they also display, on the other hand, the basic understanding (at least in it major steps) that engineers and infrastructure owners need to have while dealing with these unique and "ongoing" mechanisms in concrete infrastructure.

Over time and with the increase in technology, a wide range of techniques and devices aiming to assess the condition of concrete from both "materials" and "structures" scales have been developed; however, most of these "traditional methods", although quite promising for evaluating distress mechanisms coming from the outside to the inside of concrete (e.g., steel corrosion triggered by carbonation and or chloride penetration, external sulphate attack, or even static or dynamic loads), do not seem to be suitable to appraise the extent of internal deterioration caused by ISRs. In this context, Chapter 3 clearly points out the differences between external and internal damage mechanisms and highlights the need for a more comprehensive evaluation of the current (i.e., diagnosis: understanding the cause(s) and extent of deterioration) and future (i.e., prognosis: the possibility of further deterioration over time) condition whenever concrete structures are affected by ISRs.

Chapters 4 through 6 present various visual, non-destructive testing, microscopic and mechanical test protocols aiming to contribute towards ISRs diagnosis in concrete along with the comprehension of their impact on the engineering properties of the affected material. Although research and further developments/improvements are never-ending processes, it is widely accepted by the ISR scientific community that the current tools to diagnose and assess the condition of ISR-affected concrete are suitable for this purpose. On this basis, Chapter 7 demonstrates the use of a novel approach, the

DOI: 10.1201/9781003188155-10

so-called multi-level assessment protocol, that showed to be quite promising in assessing the cause and extent of damage of ISR-affected concrete. Chapter 9 displays a successful condition assessment campaign performed with the use of the multi-level assessment protocol to appraise a concrete overpass located in Quebec City, Canada, after nearly 50 years of service. Otherwise, the same cannot be said for the "prognosis" and "management" aspects of ISR-related research, where very little has been developed over the recent years. Chapter 8 demonstrates quite clearly that the physical and chemical test procedures developed in the laboratory to assess the potential for further development of ISRs in concrete are still preliminary and incomplete, being either qualitative or partially quantitative, where the use of the test outcomes to reassess the behaviour of affected structures is yet to be understood or implemented. Likewise, management protocols are very descriptive and heavily rely on the experience and expertise of the engineer(s) conducting the evaluation. Finally, mathematical models, although quite interesting and promising, require laboratory or field data to be adjusted. Since the laboratory tools are yet to be developed and most of the structures do not have consistent monitoring, the suitability of using modelling to forecast behaviour of ISR-affected concrete is limited. The lack of knowledge on "prognosis" aspects of ISRs makes the rehabilitation strategies and maintenance protocols adopted in practice purely experience-related, based upon past trial-and-error experiences.

The previous scenario clearly states the progress made over the last decades on the understanding of ISRs and their implication in concrete, along with tools to assess the condition of deteriorated infrastructure. However, it also emphasizes quite important challenges that engineers and researchers currently face to cope with ISRs in practice, particularly related to the potential of further deterioration of the ongoing established mechanisms. Recent developments in new technology, such as the use of high-quality sensors and imaging, coupled with artificial intelligence, etc., could improve the state-of-the-art of "prognosis" and "management" aspects of ISR research and are nowadays hot topics in the area, bearing new opportunities for further development.

Index

Pages in *italics* refer to figures and pages in **bold** refer to tables.

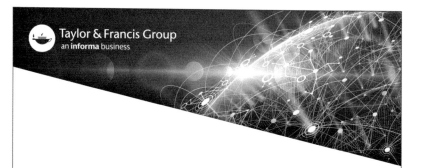